CAMBRIDGE LIBRARY COLLECTION

Books of enduring scholarly value

Zoology

Until the nineteenth century, the investigation of natural phenomena, plants and animals was considered either the preserve of elite scholars or a pastime for the leisured upper classes. As increasing academic rigour and systematisation was brought to the study of 'natural history', its subdisciplines were adopted into university curricula, and learned societies (such as the London Zoological Society, founded in 1826) were established to support research in these areas. These developments are reflected in the books reissued in this series, which describe the anatomy and characteristics of animals ranging from invertebrates to polar bears, fish to birds, in habitats from Arctic North America to the tropical forests of Malaysia. By the middle of the nineteenth century, this work and developments in research on fossils had resulted in the formulation of the theory of evolution.

Memoirs of Baron Cuvier

The author and artist Sarah Lee (1791–1856) was a remarkable traveller and scientist in her own right. With her first husband, Thomas Bowdich, she explored the flora, fauna and local culture of the Asante region and Gabon in west Africa. The pair then went in 1819 to Paris, to study zoology under the famous Cuvier, in preparation for another trip to Africa. They financed their stay by translating French scientific books into English, and became close friends of Cuvier himself. Their second expedition proved a disaster, as Thomas Bowdich died in 1824, leaving Sarah alone in Africa with three small children. She made her way back to England, and made her living from scientific translation and her own writings. This biography of her mentor was published in 1833, shortly after his death, and remained the authoritative work in English on the most distinguished scientist of the age.

Cambridge University Press has long been a pioneer in the reissuing of out-of-print titles from its own backlist, producing digital reprints of books that are still sought after by scholars and students but could not be reprinted economically using traditional technology. The Cambridge Library Collection extends this activity to a wider range of books which are still of importance to researchers and professionals, either for the source material they contain, or as landmarks in the history of their academic discipline.

Drawing from the world-renowned collections in the Cambridge University Library and other partner libraries, and guided by the advice of experts in each subject area, Cambridge University Press is using state-of-the-art scanning machines in its own Printing House to capture the content of each book selected for inclusion. The files are processed to give a consistently clear, crisp image, and the books finished to the high quality standard for which the Press is recognised around the world. The latest print-on-demand technology ensures that the books will remain available indefinitely, and that orders for single or multiple copies can quickly be supplied.

The Cambridge Library Collection brings back to life books of enduring scholarly value (including out-of-copyright works originally issued by other publishers) across a wide range of disciplines in the humanities and social sciences and in science and technology.

Memoirs of Baron Cuvier

Sarah Lee

CAMBRIDGE
UNIVERSITY PRESS

University Printing House, Cambridge, CB2 8BS, United Kingdom

Cambridge University Press is part of the University of Cambridge.
It furthers the University's mission by disseminating knowledge in the pursuit of
education, learning and research at the highest international levels of excellence.

www.cambridge.org
Information on this title: www.cambridge.org/9781108072298

© in this compilation Cambridge University Press 2014

This edition first published 1833
This digitally printed version 2014

ISBN 978-1-108-07229-8 Paperback

Drawn on Stone by C.Baugniet.

Printed by C.Motte à Londres

CUVIER.

MEMOIRS

OF

BARON CUVIER.

BY

MRS. R. LEE

(FORMERLY MRS. T. ED. BOWDICH).

LONDON:

PRINTED FOR

LONGMAN, REES, ORME, BROWN, GREEN, & LONGMAN,

PATERNOSTER-ROW.

1833.

Mon cher ami

Vous avez fort bien fait d'aller à Leyde,
puisque vous y recueillerez de nouveaux
matériaux; d'ailleurs en ce moment
vous n'auriez qu'un spectacle de désolation.
Ma pauvre fille est bien malade, et
l'inquiétude et le chagrin me tourmentent
trop pour que je puisse me livrer à aucun
travail suivi. Prenez garde aux
fièvres d'automne. Faites mes compliments
et mes remerciements à Mr. Temminck.
Adieu.

MEMOIRS

OF

BARON CUVIER.

INTRODUCTION.

Before I enter upon the subject of this volume,
I would explain to my readers the motives
which have induced me to write it, in order to
prevent that appearance of presumption, which
may naturally be laid to the charge of an un-
learned person, who attempts to write the life
of so illustrious a savant.

When death has torn from us those whom we
have most loved and revered, and the over-
whelming bitterness of grief is past, the first
feeling which awakens us from our sorrow is
the desire to uphold the memory, and to make
known to all men the virtues of the being en-
shrined in our hearts; a feeling which springs,
not only from an honest pride in doing justice

to one who is no more, but from a desire that
posterity should benefit by the example.   Rous-
ing myself, then, from the stunning grief which
at first assailed me, I eagerly sought all the
public notices which appeared in England concerning the Baron Cuvier, in the hope of finding
something equal to his high deserts; but though
all did him the justice of placing him above
every other naturalist, not one spoke of his
talents as a legislator, and all equally neglected
his private character.   This, and the almost universal incorrectness of detail, no doubt proceeded
from ignorance rather than intention; yet, disappointed as I was that my countrymen should
have so little known and appreciated one of the
most admirable persons of our time, nothing, at
that moment, could be further from my thoughts
than to supply the deficiencies by my own pen.

Most of those who were either anxious to
enquire of me concerning the surviving family, or
who were kindly solicitous about myself under
such a calamity, seemed to think it a matter of
course that I should publish some particulars of
my lost friend; but although this certainly suggested the possibility of doing so, I still felt my
own inadequacy too deeply to do other than
refuse the undertaking.   In a few weeks, how-

ever, I was solicited in one or two influential quarters to write a short memoir for one of our public journals, and, afraid to trust solely to my own reminiscences, I applied to the relatives of Baron Cuvier for data. These data were contributed with a readiness which vouched for the sentiments of the family, and I seriously applied to the task. Recollection crowded upon recollection, anecdote upon anecdote, till, in a short time, it became very difficult to select from the mass. Long did I hesitate from the conviction of my own inability; but the universal desire expressed to me that I should publish the documents which abundantly flowed from the best sources, and the anxiety evinced to know something of the private character and domestic habits of the great man, seemed to point out that part of his career which alone I was worthy to describe. Reflection whispered, that I was able to correct the many errors afloat; that, perhaps, I was the only one in England, who, from having been received into the bosom of his family, could personally speak of various circumstances and events; and when I thought of all the affection and kindness I had received, I began to feel that there would be a degree of ingratitude in remaining silent, and determined that I would,

independent of all other publications, attempt to
lay open to the English world the noblest part
of the gifted individual — his heart.

Such is the chief purport of the present biography; the labours of M. Cuvier speak for his
wonderful mind; and time alone can show, to
its full extent, the influence of that mind upon
science. To time also must we look for an impartial opinion upon his political career; but it
is only for those who have lived with him to do
justice to his high moral virtues; and in the
hope that this little volume may serve, when I
have followed the illustrious subject of it to the
grave, as a basis for a more extended publication, I offer a narrative of facts.

Having thus, I trust, obviated every feeling
of disgust which accompanies all kinds of presumption, and which would, most probably, be
attached to me, were I to dare to think myself
qualified for a biographer of savants, there yet
remains something for me to say to those to
whom I am unknown; for, when an individual
starts from a private circle to give an account of
an illustrious public character, it becomes necessary to vouch for the veracity of details, and
to explain the opportunities afforded for observation. This will be best done by a short history

of my intercourse with the Cuvier family, an introduction to whom took place through our mutually cherished friend, Dr. Leach, of the British Museum.

Mr. Bowdich had returned from his second, and I from my first, voyage to Africa, in the year 1818, and shortly after Mr. Bowdich proceeded to Paris, where his reputation, as the successful African traveller, was already known. The letter of Dr. Leach was scarcely necessary with the Baron Cuvier, who received him with that warmth and encouragement which always marked his conduct towards men of talents younger than himself, that interest which he extended to all who were devoted to science. Struck with the facilities afforded for study in the French capital, Mr. Bowdich determined to remain there some time, in order to qualify himself for the principal object of his ambition, a second travel in Africa. We both accordingly went to Paris in 1819; and from that moment the vast library of the Baron Cuvier, his drawings, his collections, were open to our purposes. We became the intimates of the family, with whom, for nearly four years, we were in daily intercourse. We left France with their blessings; and on returning alone to Europe, I was received, even as a daughter. My correspondence with M. Cuvier's daughter-in-law, and other branches of the family, has been uninterrupted since that period; I have paid them repeated visits at their own house ; and for fourteen years not a single shadow has passed over the warm affection which has characterised our intimacy.

And now, having stated my motives, and my claims to confidence, I have to express a sincere gratitude towards those who have assisted me, either by their notes or their works *, and to give an outline of the plan I have thought it necessary to adopt.

Unwilling to incur the risk of confusion, by mingling too much anecdote, either with my narrative of events or description of scientific and legislative labours, I have divided the present volume into four parts or portions, that each may bear its own share of detail. The first will give the data of all the important circumstances of the Baron Cuvier's life, in their respective order; the second will contain an account of his various works, as a savant and philosopher; the third will be devoted to his

* Foremost among these are, Baron Pasquier, M. Laurillard, Dr. Duvernoy, and the Baron de H———.

legislative career; and the fourth will be chiefly confined to those anecdotes which will best illustrate his character as a man. In following this method, I may, probably, be led into something like repetition; but I hope I shall be excused, if each part shall be found to contain a whole in itself, which facilitates reference.

PART I.

Geoʀɢᴇ Léᴏᴘᴏʟᴅ Cʜʀéᴛɪᴇɴ Fʀéᴅéʀɪᴄ Dᴀɢᴏ-
ʙᴇʀᴛ Cᴜᴠɪᴇʀ was born at Montbéliard (département du Doubs) on the twenty-third of August,
1769. This town now belongs to France, but at
that time formed a part of the kingdom of Würtemberg. His family came originally from a village of the Jura, which still bears the name of
Cuvier, and settled at Montbéliard at the period
of the Reformation. The grandfather of the subject of the present biography had two sons; one
became celebrated for his learning, and the
other, the father of George Cuvier, entered a
Swiss regiment then in the service of France.
Having much distinguished himself in his military duties, he was made Chevalier de l'Ordre
du Mérite Militaire*, which, among the Protest-
ants, was equal to the catholic Croix de St.

* The impossibility of finding English words equivalent to
French technical terms, names of public functions, orders,
&c. obliges me, in most cases, to preserve the original
phrase.

Louis; and, after forty years service, he retired, with a small pension, to Montbéliard,
where he was afterwards appointed commandant
of the artillery in that town. At fifty years of
age he married a young lady, gifted with much
talent and feeling, by whom he had three sons.
The eldest died while his mother was pregnant
with her second son, which event preyed so
much upon her health, that her infant, George,
came into the world with a constitution so
feeble, that his youth scarcely promised man-
hood. The cares of this excellent mother,
during the extreme delicacy of his health, left
an impression on M. Cuvier which was never
effaced, even in his latest years, and amid the
absorbing occupations of his active life. He
cherished every circumstance connected with
her memory; he loved to recall her kindnesses,
and to dwell upon objects, however trifling,
which reminded him of her. Among other
things, he delighted in being surrounded by
the flowers she had preferred, and whoever
placed a bouquet of red stocks in his study or
his room, was sure to be rewarded by his
most affectionate thanks for bringing him what
he called " the favourite flower." But this
well-judging parent did not confine her cares

to his health alone ; she devoted herself equally
to the formation of his mind, and was another
proof of the influence that a mother's early attentions frequently shed over the future career
of her son. She guided him in his religious
duties, taught him to read fluently at the age of
four years, took him every morning to an elementary school, and, although herself ignorant
of Latin, so scrupulously made him repeat his
lessons to her, that he was always better prepared with his tasks than any other boy at the
school. She made him draw under her own
inspection ; and, by constantly furnishing him
with the best works on history and general literature, nurtured that passion for reading, that
ardent desire for knowledge, which became the
principal spring of his intellectual existence. As
he advanced in drawing, his progress was super-
intended by one of his relations, an architect in
the town of Montbéliard; and he successively
passed through all the exercises of this first
school, repeating the usual catechisms, the
psalms of David, and the sonnets of Drelin-
court, &c., with the utmost facility. At ten
years of age he was placed in a higher school,
called the Gymnase, where, in the space of four
years, he profited by every branch of education

there taught, even including rhetoric. He had
no difficulty in acquiring Latin and Greek, and
he was constantly at the head of the classes of
history, geography, and mathematics. The history of mankind was, from the earliest period of
his life, a subject of the most indefatigable application; and long lists of sovereigns, princes,
and the driest chronological facts, once arranged
in his memory, were never forgotten. He also
delighted in reducing maps to a very small
scale, which, when done, were given to his companions; and his love of reading was so great,
that his mother, fearing the effect of so much
application to sedentary pursuits, frequently
forced him to seek other employments. When
thus driven, as it were, from study, he entered
into boyish sports with equal ardour, and was
foremost in all youthful recreations. It was at
this age that his taste for natural history was
brought to light by the sight of a Gesner, with
coloured plates, in the library of the Gymnase,
and by the frequent visits which he paid at the
house of a relation who possessed a complete
copy of Buffon. Blessed with a memory that
retained every thing he saw and read, and which
never failed him in any part of his career, when
twelve years old he was as familiar with qua-

drupeds and birds as a first-rate naturalist.   He
copied the plates of the above work, and coloured
them according to the printed descriptions,
either with paint or pieces of silk.   He was never
without a volume of this author in his pocket,
which was read again and again; and frequently
he was roused from its pages to take his place
in the class repeating Cicero and Virgil.   The
admiration which he felt at this youthful period
for his great predecessor never ceased, and in
public, as well as private circles, he never failed
to express it.   The charms of Buffon's style, a
beauty to which M. Cuvier was very sensible,
had always afforded him the highest pleasure,
and he felt a sort of gratitude to him, not only
for the great zeal he had evinced in the cause of
natural history, not only for the enjoyment
afforded to his youthful leisure, but for the
many proselytes who had been attracted by the
magic of his language.   When the student had
ripened into the great master, M. Cuvier found
me deeply absorbed by a passage of Buffon; and
he then told me what his own feelings had been
on first reading him, and that this impression had
never been destroyed in maturer years.   He had
been obliged, for the sake of science, to point
out the errors committed by this eloquent na-

turalist, but he had never lost an opportunity of
remarking and dwelling on his perfections.

At the age of fourteen we find the dawning
talents of the legislator manifesting themselves;
and the young Cuvier then chose a certain
number of his schoolfellows, and constituted
them into an academy, of which he was appointed president. He gave the regulations,
and fixed the meetings for every Thursday, at
a stated hour, and, seated on his bed, and
placing his companions round a table, he ordered that some work should be read, which
treated either of natural history, philosophy,
history, or travels. The merits of the book
were then discussed, after which, the youthful
president summed up the whole, and pronounced
a sort of judgment on the matter contained in
it, which judgment was always strictly adopted
by his disciples. He was even then remarkable
for his declamatory powers, and on the anni-
versary fête of the sovereign of Montbéliard,
Duke Charles of Würtemberg, he composed an
oration in verse, on the prosperous state of the
principality, and delivered it fresh from his
pen, in a firm manly tone, which astonished
the whole audience. Like most of the young
people at Montbéliard, whose talents rendered

them worthy of it, and whose parents were not
possessed of fortune, he was destined for the
church. A free school had been founded for
such boys at Tubingen, where they received
a first-rate education. But the chief of the
Gymnase at Montbéliard, who had never forgiven the young Cuvier for some childish
tricks, changed his destiny by placing his composition in the third rank, when the pupils
presented their themes for places. George Cuvier felt that his production was equally good
with those which had hitherto been judged
worthy of the first rank, and at the important
moment, when his station at college depended
on his success, he was, for no conscious fault, kept
back. He became disgusted, and abandoned
all thoughts of Tubingen, to which place he
was only desirous of going as a means of pursuing his studies; and, frequently, in after-life,
he expressed himself most happy at the changes
which resulted from this piece of injustice.

Informed of the progress of the young Cuvier, and hearing the highest encomiums of him
from the Princess his sister, the Duke Charles,
uncle to the present King of Würtemberg,
when on a visit to Montbéliard, sent for him,
and, after having asked him several questions,

and examined his drawings, declared his intention of taking him under his special favour, and
sending him to the University of Stuttgard free
of expense, there to enter into his own Academy,
called the Académie Caroline. He was then
only fourteen, but, in consequence of the preparation he had undergone at the Gymnase of
Montbéliard, he was able to take his place
among the most celebrated students of the Academy. He, at this age then, quitted the paternal roof for the first time : he was sent among
strangers without having an idea of the establishment he was about to enter ; and even in his
latest years he often said, that he could not
recall to memory this three days journey without a sensation of fear. He was seated between
the Chamberlain and Secretary of the Duke,
both entirely unknown to him, and who spoke
nothing but German the whole way, of which
the poor child could not understand one word.
On the 4th of May, 1784, he entered the Académie Caroline ; and during the four years he
passed there, he studied all that was taught in
the highest classes, — mathematics, law, medicine, administration, tactics, commerce, &c.
After applying himself for one year to philosophy, as his particular object, he then chose the

study of administration, which, in Germany, embraces the practical and elementary parts of law, finance, police, agriculture, technology *, &c., and was principally led to this preference, because it also afforded him many opportunities of pursuing natural history, of herborising, and of visiting collections. He, on all occasions, enthusiastically profited by these opportunities, for the cultivation of his darling taste; he frequently read over Linnæus, Reinhart, Mur, and Fabricius. In his walks he collected a very considerable herbarium; and, during his hours of recreation, he drew and coloured an immense number of insects, birds, and plants, with the most surprising correctness and fidelity, and to which drawings he would frequently return with pleasure, when the naturalist was perfect in his career. But it was the same in every thing; for that versatility of talent, which made him the wonder of all who knew him as a man, seems to have distinguished him in early years. He obtained various prizes, and the order of Cheva-

* Technology is the theoretical part of mechanical science, independent of the practical; a knowledge of which was thought absolutely indispensable to one taking a part in administration.

lerie *, — an honour which was only granted to five or six out of four hundred pupils; and nine months after his arrival at Stuttgard, he bore off the prize for the German language.

The youthful Cuvier was destined solely to fill the higher departments belonging to the government of his country; but the pecuniary embarrassments of his parents rendered it impossible for him to wait two or three years, till an opportunity of appointing him should occur to the Duke. The disordered state of the finances in France was so great, that even the payment of his father's pension had ceased, and he was consequently forced to enter into a career wholly different to his own wishes, or to the views of his patron. Duke Frederick, who was governor of Montbéliard, under his brother, Duke Charles, retired to Germany, and in him M. Cuvier lost one of his most able protectors; and every

* The chevaliers dined at a separate table, and enjoyed many advantages, as being under the immediate patronage of the Duke. The lessons of M. Kielmeyer, afterwards called the father of the philosophy of nature, a student much older than himself, were of infinite service to M. Cuvier at this time, as from him he learned to dissect, and with him, Messrs. Pfaff, Marschall, Hartmann, &c., a society of natural history was formed; and he who brought the best composition to the meetings received an order, beautifully drawn by M. Cuvier.

hope of better times failing, he determined to
undertake the office of tutor, an idea in some
measure familiar to him, as Montbéliard had
long supplied instructors to the young nobles of
Russia. To Russia, however, he had no wish to
proceed, for his lungs, always delicate, were
rendered still weaker by close application to his
studies, and he sought an appointment in a more
genial climate. Such a step was deemed by his
companions, considering his already acquired
honours, his extraordinary talents, and great attainments, desperate; but he was again to prove,
that that which at first appears a severe misfortune
often becomes a stepping stone to future fame
and success; for, in a manner compelled to accept that which in every way appeared unworthy of him, M. Cuvier, by so doing, laid the
foundation for the cosmopolitan honours which
attended his after years. We are now to behold
him, then, arriving at Caen in Normandy, in
July, 1788, and stationing himself in a Protestant
family for the education of the only son, and
although not quite nineteen years of age, in possession of that variety and depth of knowledge
which was so soon to ripen into the great savant;
" bringing with him from Germany that love of
labour, that depth of reflection, that persever-

ance, that uprightness of character, from which
he never swerved. To these admirable found-
ations for glory, he afterwards added that remarkable clearness of system, that perfection
of method, that tact of giving only what is
necessary, in short, that elegant manner of summing up the whole, which particularly distinguishes the French writers : the whole super-
structure was completed by the most perfect
modesty, and that respect for his own esteem,
without which, talents become the medium of
traffic for the acquirement of sordid possessions."*

Whilst with the family of the Count d'Hericy,
M. Cuvier saw all the nobility of the surrounding country; he acquired the forms and manners
of the best society, and became acquainted with
some of the most remarkable men of his time.
Nor was his favourite study followed with
less ardour in consequence of finding himself
surrounded by new friends and new duties. A
long sojourn on the borders of the sea first induced him to study marine animals, but, without
books, and in complete retirement, he confined
himself to the objects more immediately within
his reach. It was at this period also, (June,

* Baron Pasquier.

1791, to 1794,) that some Terebratulæ having
been dug up near Fécamp, the thought struck
him of comparing fossil with recent species*;
and the casual dissection of a Calmar† led him to
study the anatomy of Mollusca, which afterwards
conducted him to the developement of his great
views on the whole of the animal kingdom. It
was thus, from an obscure corner of Normandy,
that that voice was first heard, which, in a comparatively short space of time, filled the whole
of the civilised world with admiration, —which
was to lay before mankind so many of the
hidden wonders of creation, —which was to discover to us the relics of former ages, to change
the entire face of natural history, to regulate
and amass the treasures already acquired, and
those made known during his life; and then to
leave science on the threshold of a new epocha.
The class called Vermes by Linnæus, included
all the inferior animals, and was left by him in a
state of the greatest confusion. It was by these,

* The idea of making fossil remains subservient to geology was not due to M. Cuvier alone, for several others seem
to have entertained the same views; but his pre-eminence
consisted in making use of this idea, and carrying it to an
extent far beyond the calculations of his predecessors or
contemporaries.
† A species of Cuttle fish.

the lowest beings in creation, that the young naturalist first distinguished himself : he examined
their organisation, classed them into different
groups, and arranged them according to their
natural affinities. He committed his observations and thoughts to paper, and, unknown to
himself at that time, laid the basis of that beautiful fabric which he afterwards raised on zoology. He wrote concerning them, to a friend,
" These manuscripts are solely for my own use,
and, doubtless, contain nothing but what has
beer done elsewhere, and better established by
the naturalists of the capital, for they have been
made without the aid of books or collections."
Nevertheless, almost every page of these precious manuscripts was full of new facts and enlightened views, which were superior to almost all
that had yet appeared. A little society met every
evening in the town of Valmont, near the château de Fiquainville, belonging to the Count
d'Hericy, for the purpose of discussing agricultural topics. M. Tessier was often present
at these meetings, who had fled from the reign
of terror in Paris, and who was concealed under
the title and office of surgeon to a regiment,
then quartered at Valmont. He spoke so well,
and seemed so entirely master of the subject,

that the young secretary of the society, M. Cuvier, recognised him as the author of the articles on agriculture in the Encyclopédie Méthodique.

On saluting him as such, M. Tessier, whose title of Abbé had rendered him suspected at Paris, exclaimed, " I am known, then, and consequently lost."— " Lost!" replied M. Cuvier; "no; you are henceforth the object of our most anxious care." This circumstance led to an intimacy between the two; and by means of M. Tessier*, M. Cuvier entered into correspondence with several savans, to whom he sent his observations, especially Laméthrie, Olivier, De la Cépéde, Geoffroy St. Hilaire, and Millin de Grand Maison. Through their influence, and from the memoirs published in several learned journals, he was called to Paris, where endeavours were making to re-establish the literary institutions, overthrown by the Revolution, and where it was reasonable to suppose that he would find the means of placing himself. In the spring of 1795, he obeyed the invitation

* " Je viens de trouver une perle dans le fumier de Normandie,"—" I have just found a pearl in the dunghill of Normandy,"— wrote M. Tessier to his friend M. Parmentier ; thus detecting the great naturalist in M. Cuvier's earliest productions, and appreciating what were then but the germs of his talent.

of his Parisian friends, and, by the influence of
M. Millin, was appointed membre de la Commission des Arts, and, a short time after, professor
at the central school of the Panthéon. For this
school he composed his "Tableau élémentaire
de l'Histoire naturelle des Animaux;" which
work contained the first methodical writing on
the class Vermes that had been given to the
world. His great desire, however, was to be
attached to the Museum of Natural History, the
collections in which could alone enable him to
realise his scientific views. A short time after his
arrival in the capital, M. Mertrud was appointed
to the newly-created chair of comparative anatomy at the Jardin des Plantes, and finding
himself too far advanced in years to follow a
study which had hitherto been foreign to his
pursuits, consented, at the request of his colleagues, particularly MM. de Jussieu, Geoffroy,
and De la Cépéde, to associate M. Cuvier with
him in his duties. This association was exactly
what M. Cuvier was desirous of obtaining; and
no sooner was he settled in the Jardin des
Plantes, as the assistant of M. Mertrud, July,
1795, than he sent for his father, then nearly
eighty years of age, and his brother, M. Frederic Cuvier; his mother he had unfortunately

lost in 1793. From the moment of his instal-
lation in this new office, M. Cuvier commenced
that magnificent collection of comparative anatomy which is now so generally celebrated. In
the lumber-room of the museum were four or
five old skeletons, collected by M. Daubenton,
and piled up there by M. de Buffon. Taking
these, as it were, for the foundation, he unceasingly pursued his object; and, aided by some
professors, opposed by others, he soon gave it
such a degree of importance that no further
obstacle could be raised against its progress.
No other pursuit, no relaxation, no absence, no
legislative duties, no sorrow, no illness, ever
turned him from this great purpose, and created
by him, it now remains one of the noblest monuments to his memory.*

The National Institute was created in 1796;
and M. Cuvier, although only known by his
scientific papers, and his intimacy with learned
men, especially De la Cépéde and Daubenton,
was made one of its first members, and was the

___

* It was of this collection that he said, when asked if he
should ever consider himself rich in it, " Quelque riche qu'on
en soit, on en désire toujours." (However rich we may be,
we always wish for more.)

third secretary, appointed at a time when these secretaries quitted their office every two years.

In the spring of 1798, M. Berthollet having been charged by Buonaparte to seek for savans to accompany the expedition to Egypt, proposed to M. Cuvier to form one of the number. This, however, he refused, from the conviction, that he could better serve the interests of science by remaining amid the daily improving collections of the Jardin, where his labours could be systematic, than by making even a successful travel. He always felt happy afterwards in having thus decided; the propriety of which resolution no one can attempt to dispute.

About this time, one of M. Cuvier's pupils, M. Dumeril, who had zealously followed all his lectures, asked permission to publish the notes he had taken in the lecture room. These, in M. Cuvier's opinion, would have formed a very imperfect work, and he preferred going over the whole again, devoting himself to the general and philosophical notices, and those parts which treated of the brain and the organs of the senses. M. Dumeril chiefly undertook the details of myology and nevrology. The two first volumes of the " Leçons d'Anatomie comparée" appeared in 1800, and met with the greatest success, notwithstanding a few errors, which were afterwards corrected and acknowledged by M. Cuvier, who, in common with all those who prefer the interests of science to their own momentary fame, and with the candour which always marks real learning, never hesitated either to avow or to rectify a fault, a perfection which mingled with his private as well as public actions. The materials for these lectures were supplied by a collection, then in its infancy, and which was increased an hundredfold by himself; and those who have criticised these early volumes, have been obliged to confess, that the means of doing so were given to them by the author himself, who threw every thing open to them, even were it to convict him of those unavoidable mistakes to which he had been liable, from the then imperfect state of the collection. The three last volumes of this work were much more complete and methodical than the first two, and were edited under the inspection of Dr. Duvernoy (another of M. Cuvier's pupils), in the year 1805, though the second, notwithstanding its inaccuracies, was always considered by M. Cuvier as the most interesting of the whole.

But to return to the year 1800, when the celebrated colleague of M. de Buffon died, at a

very advanced age, M. Cuvier was named professor in his place, at the Collège de France,
where he taught natural philosophy, at the same
time that he lectured on comparative anatomy
at the Jardin.* On succeeding to this chair he
resigned that of the central school of the Panthéon. Also in 1800, Buonaparte, who, as First
Consul, aspired to civil as well as military glory,

---

* An estimate of the pecuniary advantages then attending
the career of a savant, may be gathered from the following
letter, written by M. Cuvier, in answer to one from the late
M. Hermann.

" My dear and learned confrère,   (1800.)
" You are not to suppose that Paris is so highly favoured;
for twelve months pay are now due at the Jardin des
Plantes, and all the national establishments for public instruction, in Paris as well as at Strasburgh; and if we envy
the elephants, it is not because they are better paid than we
are, but because, while living on credit, as we do, they are
not aware of it, and, consequently, are insensible to the pain
it gives. You know the saying about the French, that when
they have no money they sing. We savans, who are not
musicians, work at our sciences instead of singing, which
comes to the same thing. Believe me, my dear confrère,
this French philosophy is better than that of Wolff, or even
that of Kant; and you are even more able to profit by it
than we are, for you can still purchase beautiful books, and
even artificial anatomy, which are objects of luxury in their
way. I have not yet read Poli, and defer this study till the
time when I publish my anatomical history of animals with
white blood. There is, as yet, but one copy of it in Paris, as
I am informed; and thus you see we offer nothing which can
excite your envy."

caused himself to be appointed president of the
Institute, and, in consequence, held direct com-
munication with M. Cuvier. In 1802 he appointed him one of the six inspectors-general
ordered to establish Lycées* in thirty towns of
France. In this capacity M. Cuvier founded
those of Marseilles, Nice, and Bordeaux, which
are now called royal colleges; and while thus
employed at Marseilles, he profited by the opportunity so afforded him of continuing his
studies on marine animals. During his absence
from Paris, the Institute underwent a change
of form, and its secretaryships were made
perpetual. †   M. Cuvier was elected to that
of natural sciences, which he held with honour to the day of his death. On this ap-

---

* Lycées are public schools, under the management and
direction of the government. The pupils who frequent
them pay a small sum, which sum is appropriated to the use
of the school. The professors receive their salaries from the
government, which reserves to itself a right to nominate a
certain number of pupils entirely gratis. The private
schools are always established near one of these Lycées, as
the pupils of these are obliged to attend there for a certain
number of hours every day.

† Napoleon fixed the salary of the perpetual secretaries of
the Institute at 6000 francs; and on its being observed to
him that it was too much, he replied, " The perpetual
secretary must be enabled to receive at dinner all the
learned foreigners who visit the capital."

pointment he quitted his labours of inspector-
general of education.

A fall having occasioned the death of M.
Cuvier's father, shortly after his arrival in Paris,
and his brother's wife having died the first year
of her marriage, in giving birth to a son*, the
two brothers remained alone ; and it was in this
comparatively solitary condition that M. Cuvier
thought of seeking a companion. In 1803 he married the widow of M. Duvaucel, Fermier General,
who had perished on the scaffold in the year
1794. This was no match of interest; for
Madame Duvaucel had been wholly deprived of
fortune by the Revolution, and brought four
children † to M. Cuvier, whom she had borne to

* M. Frederic Cuvier is now keeper of the ménagerie of
the Jardin des Plantes, in which capacity his observations on
the instinct, habits, and dentition of animals have been
highly valuable.   He is the author of several learned works
on these subjects, is member of the Institute, one of the inspectors-general of education, &c. &c. ; but all these titles to
public consideration are nothing in comparison to the admirable qualities of his heart and temper.   The distinguished
talents of the son thus bequeathed to him will at least bear
the illustrious name of Cuvier one generation further with
honour.

† Two of these children are dead, one of them having
been assassinated in Portugal during the retreat of the
French in 1809.   The other fell a victim to his scientific
zeal in a pernicious climate ; and after having displayed great

M. Duvaucel. But well had M. Cuvier judged
of the best means of securing domestic enjoyment; for this lady, who is a rare combination
of mind, manner, and disposition, threw a bright
halo of happiness round him, which was his support in suffering, his refuge in trouble, and a
powerful auxiliary, when his heavy and important duties allowed him to steal an hour of rational and unrestrained conversation. By this
marriage he had four children, the first of whom,
a son, died a few weeks after his birth, and who
were all successively taken from him.

In 1808, in his character of Perpetual Secretary, M. Cuvier wrote a Report on the Progress
of Natural Sciences, from the year 1789. A
mere report was demanded; but under this title
the learned author produced one of the most
luminous treatises that had ever appeared,

---

talent and courage, while travelling in India and the neighbouring islands for four years, in order to make collections for the museum in Paris, expired at Madras, at an
early age, lamented by all as a youth of great promise, and
the most endearing qualities. One of the survivors holds a
high place in the customs of Bordeaux; and the other, who
has been loved and cherished by M. Cuvier as his own
daughter, has had the happiness of devoting herself to him
in his last moments, and now forms the sole consolation of
her afflicted mother.

" serving as a beacon to the path which had already been traversed, and to that which was yet to be pursued."* The Report was formally presented to the Emperor in the council of state. In this same year, when Napoleon created the Imperial University, M. Cuvier was made one of the counsellors for life to this body, which brought him constantly into the immediate presence of the Emperor.

In 1809 and 1810, in his office of Counsellor to the University, M. Cuvier was charged with the organisation of the academies of those Italian states which were, for a time, annexed to the empire. The regulations made by him at Turin, Genoa, and Pisa, were afterwards continued by the sovereigns of these countries on their return to their dominions.

In 1811 appeared one of the most important of all M. Cuvier's scientific labours, — his work on Fossil Remains; which opened new sources of wonder in the history of creation, and made an entire revolution in the study of geology. Also, in 1811, he was ordered to form academies in Holland and the Hanseatic towns,

* Baron Pasquier.

where several of his arrangements are still existing. His Reports from Holland are particularly worthy of admiration; for in them he exposed the true causes of the inferiority of that country in classical attainment, and showed, that the disgust often felt by the pupils, arose from their not having enough given to their minds to feed upon. The schools for the people attracted his attention in all countries, and were to him an unceasing theme of meditation.

While at Hamburgh, M. Cuvier received the unsolicited title of Chevalier from the Emperor, which rank was assured to his heirs. However, the hope of transmitting his worldly honours to his posterity was soon to be destroyed; for, after being deprived of a daughter, four years old, in 1812, he was, in 1813, bereaved of his son, who was seven years of age. This last loss made a deep impression on him, which was never entirely effaced; and even after the lapse of years he never saw a boy of that age without considerable emotion, a feeling which he did not strive to hide from his own family, or those with whom he was intimate; and often, when walking with his daughters, he would stop before a group of boys, who, as they played, reminded him of his child.* This misfortune happened while M. Cuvier was fulfilling a mission at Rome, for the purpose of organising the university there. It was remarkable enough, that a Protestant should hold this office in the metropolis of the Papal dominions, but the moderation and benignity of M. Cuvier knew how to soften inconsistencies; his tolerance for all sincere doctrines of religion proceeded from conscientious motives, and therefore he was not likely to revolt the creed of those among whom he mingled. While thus employed at Rome, Napoleon, from his own personal feeling, appointed him Maître des Requêtes in the Council of State, of which honour he was first informed by the Moniteur. The contact into which he was constantly brought with the Emperor, in his office of Counsellor to the University, the intimate knowledge which his sovereign had thus acquired of his administrative talents, united to the favourable representations of the Grand Master, Fontanes, were supposed to be

* So late as 1830, when M. Cuvier visited this country, I took my son to see him at the hotel where he was staying, forgetting the effect it was likely to produce; and I shall for ever remember the pause he made before him, and the melancholy tenderness with which he laid his hand on the head of the boy.

D

the causes of this marked distinction. Towards
the end of this year (1813) he was further employed by Napoleon, in a manner that showed
the estimate he had made of his character. He
appointed him Commissaire Impérial extraordinaire, and sent him on the difficult mission
of endeavouring to raise the people inhabiting
the left bank of the Rhine in favour of France,
(their new country) against the invading troops
then marching against her. M. Cuvier was
ordered to Mayence; but he was stopped at
Nancy, by the entrance of the allied armies,
and obliged to return.

The events of 1814 happened at the moment
when the Emperor had bestowed on him a still
more honourable mark of his favour, by making
him Counsellor of State. A delay of only a few
months, however, took place in his final establishment in the council; for Louis XVIII., who
was very sensible to intellectual merit, again
conferred this dignity on him, and, in the September of the same year, first employed him in
the temporary office of Commissaire du Roi.
These favours were, in some measure, to be attributed to an introduction to the Abbé de Mon-
tesquion, then minister, by means of MM. Royer
Collard, Becquey, de Talleyrand, and Louis,

who were well acquainted with the Abbé, and who, by their presentation, gave him an opportunity of profiting by the merits of M. Cuvier.

The return of Napoleon for a while banished the new counsellor from his dignity, but he was retained by the Emperor in the Imperial University. After the hurricane of the Hundred Days it became necessary to remodel both the Royal and Imperial Universities, and a provisional superintendence was deemed necessary. A committee of public instruction was created to exercise the powers formerly belonging to the grand master, the council, the chancellor, and the treasurer of the University. M. Cuvier made a part of this committee, and was at once appointed to the chancellorship, which office he retained till his death, under the most difficult circumstances, in the midst of the most opposite prejudices, and notwithstanding the most inveterate resistance offered to him as a Protestant. The jesuitical tendency of those in power augmented the difficulties that a wise and disinterested man must at all times meet with, in trying to do good, and to prevent evil; but when that man was of a different religion, it may easily be imagined in how delicate a situation he must have been often placed, and how greatly

his religious faith must have increased the obstacles he had to encounter. To those unac-
quainted with the early part of M. Cuvier's
career, it would seem extraordinary, that all
these high functions should be conferred on a
naturalist by profession, but it should be considered, that he only thus pursued his original
destination, out of which he had been thrown
by political events; that he had only changed
his master, and become counsellor of state to a
great king instead of a petty prince. From
this period he took a very active part, not precisely in political measures, properly so called,
from which he by choice withdrew himself as
much as possible, but in projects for laws, and
every sort of administration, which especially
belonged to the Committee of the Interior attached to the Council of State. He was also,
generally speaking, the Commissaire du Roi,
appointed for defending the new or ameliorated
laws before the two Chambers.

During the first years of the restoration of
the Bourbons, M. Cuvier was twice offered the
directorship for life of the Museum of Natural
History, but he persisted in refusing it, from the
conviction that it was much more favourable to
the advancement of science, that this establish-

ment should continue under that form of administration, which necessitated the election of a
yearly director, chosen by the professors, and
appointed according to their vote.   A second
edition of the Fossil Remains was published in
1817, the preliminary discourse of which underwent several more editions.   The Régne Animal was also brought out in this year, which
classed every branch of zoology according to its
organisation.   In 1818 M. Cuvier made a journey to England with his family and his secretary, the excellent M. Laurillard, and where he
remained about six weeks, visiting every thing
worthy of notice in London.   His remark to
his Majesty George IV. concerning our natural history was, that if the private collections
could be amassed into one, they would form a
great national museum, which would surpass
every other.   At this period the election for
Westminster was going forward, and he frequently dwelt on the amusement he had received from being on the hustings every day.
These orgies of liberty were then unknown in
France, and it was a curious spectacle for a man
who reflected so deeply on every thing which
passed before him, to see and hear our orators
crying out at the tops of their voices to the

mob, who pelted them with mud, cabbages, eggs, &c. ; and Sir Murray Maxwell, in his splendid uniform, and decorated with orders, flattering the crowd, who reviled him, and sent at his head all the varieties of the vegetable kingdom. Nothing ever effaced this impression from M. Cuvier's memory, who frequently described the scene with great animation.

M. Cuvier had two objects in visiting England, one of which was, to observe, on the spot, the influence of our constitutional government, which was only known to him in theory. He conversed with several of our political characters, he saw every thing which marked the application of our system upon mankind, and took back with him to France clear and precise ideas, by which he well knew how to profit in his future labours. It was frequently a matter of great astonishment to my countrymen to find him so well acquainted with our institutions, even to the details of their expenses, the period of their formation, and the changes they had undergone. The other, and the great object of M. Cuvier's excursion, was of a scientific nature; and it is with pleasure I add, that he always spoke of his reception here with gratitude. The facilities afforded him both by our savants and

our statesmen, the confidential communications
he received, and the manner in which all was
laid open to him, were frequently a source of
happy recollection, which was as often expressed.
Some days of the period of his sojourn in England were passed at Oxford, whither he was
accompanied by his valued friend, Dr. Leach of
the British Museum, who was his incessant cha-
peron in this country; he returned from thence
perfectly enchanted with the city and its great
objects of interest, and with the distinction
which attended his reception there. His wife
and daughters met him at Windsor, and, after
passing the day in visiting the castle, park, &c.,
they proceeded, late in the evening, to the
house of Sir William Herschel, who received
them with the utmost kindness, and showed
them his great telescope, though the night was
too dark to profit much by this famous instrument. Another visit paid by M. Cuvier was
often alluded to by him with pleasure; it was
to Sir Joseph Banks's house at Spring Grove:
he had often been to see him in Soho Square,
but the entertainment given to the whole party
at Spring Grove resembled a fête champêtre.
The only thing to which M. Cuvier could not

reconcile himself in England was, the formality
and length of our great dinners, the long sittings after which were always mentioned by
him with an expression of ennui, even in his
countenance. At one of these sittings, at Sir
Everard Home's, the conversation turned upon
some political question. In the course of the discussion M. Cuvier said, — " But it would be very
easy to clear up this point, if Sir Everard would
send to his library for the first volume of Black-
stone's Commentaries." Upon this Sir Everard,
with great emphasis, exclaimed, " Know, Monsieur, that I have not such a book in my library,
which, thank God, only contains works of science." To this M. Cuvier quietly replied,
" The one does not prevent the other;" but
never could recollect this, to him extraordinary
boast, without a mixture of amusement and
astonishment. While in England, M. Cuvier
was appointed to the Académie Française,
chiefly in consequence of the brilliant éloges
he had read in the Academy of Sciences on its
deceased members. His discourse upon his reception is a beautiful instance of his classical
style of writing. Towards the end of 1818 he
was offered the Ministry of the Interior, but the
political conditions attached to it being such as

he could not conscientiously accept, he declined the honour.

In 1819 M. Cuvier was appointed President of the Comité de l'Intérieur, belonging to the Council of State, an office which he held under all changes of ministry ; because, notwithstanding its importance, it is beyond the reach of political intrigue, and only demands order, unremitting activity, strict impartiality, and an exact knowledge of the laws and principles of administration. In this same year, Louis XVIII., as a mark of personal esteem, created him a Baron*, and repeatedly summoned him to assist in the cabinet councils.

Twice had M. Cuvier held the office of Grand Master of the University, when the place could not conveniently be filled up, but he never received the emoluments of it ; and, in 1822, when a Catholic bishop was raised to this dignity, he accepted the Grand Mastership of the Faculties of Protestant Theology ; on assuming which, he made conditions, that he should not receive any

* A week after M. Cuvier received this title he went to the theatre, and in the course of the evening one of the actors exclaimed, in his part, " and for all these services, the King has only created him a Baron." The audience gaily applied the sentence to M. Cuvier, who was as much amused as any of them at the coincidence.

pecuniary reward. This appointment associated
him with the ministry, and gave him the superintendence, not only of the religious, but the civil
and political rights of his own creed, and ceased
only with his life, although the Grand Masters
were afterwards laymen.

In 1824, M. Cuvier officiated, as one of the
Presidents of the Council of State, at the coronation of Charles X.; and, in 1826, received from
that monarch the decoration of Grand Officer
de la Légion d'Honneur. On the Saturday he
knew nothing of this compliment, and on Sunday
it arrived, without, however, disturbing him from
the delighted survey he was taking, with his
daughter-in-law, of some alterations just made
in his house. At this time also, his former sovereign, the King of Würtemburg, appointed him
Commander of his Order of the Crown.

In 1827, to M. Cuvier's Protestant Grand
Mastership was added the management of all
the affairs belonging to the different religions in
France, except the Catholic, in the Cabinet of
the Interior, for which increase of his duties he
also refused to accept any emolument. But this
year was marked with the heaviest calamity the
Baron Cuvier had yet sustained, the loss of
his only remaining child; a pious, talented,

beautiful young woman of twenty-two, on the
eve of marriage, and whose bridal chaplet
mingled with the funeral wreath on her bier.
Lovely in every action, lovely in person and
manner, and rich in her attainments, no question
ever arose as to who did or did not admire
Clementine Cuvier; she unconsciously com-
manded universal homage, and secured its continuance by her lowliness of heart and her unfailing charity. The daughter was worthy of
the father: it may be imagined, then, how that
father loved her, and how heavy was the visitation. But M. Cuvier, with that high sense of
duty which had always distinguished him, felt
that he lived for others, and that he had no right
to sink under the heavy load of grief imposed on
him. With the energy that might be expected
from such a character, he sought relief in his
duties; and although many a new furrow appeared on his cheek; although his beautiful hair
rapidly changed to silvery whiteness; though
the attentive observer might catch the suppressed
sigh, and the melancholy expression of the up-
lifted eye, no one of his important offices remained neglected; his scientific devotion even
increased; his numerous protégés received the
same fostering care, and he welcomed strangers

to his house with his wonted urbanity. It has been related by an eye-witness, that, at the first sitting of the Comité de l'Intérieur at which M. Cuvier presided after this event, and from which he had absented himself two months, he resumed the chair with a firm and placid expression of countenance; he listened attentively to all the discussions of those present; but when it became his turn to speak, and sum up all that had passed, his firmness abandoned him, and his first words were interrupted by tears; the great legislator gave way to the bereaved father; he bowed his head, covered his face with his hands, and was heard to sob bitterly. A respectful and profound silence reigned through the whole assembly; all present had known Clementine, and therefore all could understand and excuse this deep emotion. At length M. Cuvier raised his head, and uttered these few simple words : — " Pardon me, gentlemen; I was a father, and I have lost all;" then, with a violent effort, he resumed the business of the day with his usual perspicuity, and pronounced judgment with his ordinary calmness and justice.

In the following year (1828) appeared the first of a series of twenty volumes on Ichthyology, a magnificent work, accompanied by the most

exquisite plates. In 1829, a second edition of
the Règne Animal was published; and it is
scarcely possible to imagine any thing finer than
the force of that mind, which could thus seek
for solace under the deepest affliction. These
works were in progress long before the death of
Mademoiselle Cuvier, and, we may safely suppose,
were not much retarded by that grievous event.
What was the state of the father's mind during
the time of her illness, may be gathered from
a letter, published in the second part of this
volume.

The year 1830 saw the Baron Cuvier again
in the lecturing chair at the Collége de France,
where he opened a course on the History and
Progress of Science in all Ages, and which was
continued till the close of his earthly labours.
In the same year he paid a second visit to
England, and happened to be in London when
the last revolution in France took place. He
had long contemplated this visit, being desirous
of personally inspecting some of the scientific
treasures of this country; but a long delay
(even after his congé was obtained) took place,
owing to the death of the learned Baron Fourrier, the other secretary to the Académie des
Sciences, whose duties fell on M. Cuvier till a

successor could be appointed. On the publication of the famous ordonnances of Charles X.
and his ministers, an universal silence in public
was observed, as if the first person who ventured
to talk about them, was to set fire to a train of
gunpowder. Even M. Cuvier, though so clear-
sighted on other occasions, was completely taken
by surprise in this instance, and partook of the
general opinion, that " this stroke of policy on
the part of the state would lead to a lengthened
resistance of taxes, and to partial disturbances,
but not to any violent crisis ; " and deceived, as
so many others were, by the profound tranquillity which reigned in every part of the capital,
he started for England on the appointed day.
Five hours after his carriage had passed the barrier the firing commenced in Paris, and he and
his daughter-in-law quietly pursued their route
by easy stages. They were overtaken on the
road near Boulogne by the flying English, who
gave them vague reports, and they pressed on to
meet their letters at Calais. There, after two
days of the deepest anxiety, during which time
they formed twenty projects for immediate return, and were as often retained by the certainty
of not being able to re-enter Paris, or even proceed on the road back, with passports dated in

the month of May, and leave of absence signed
by the hand of Charles X., they at once received
the details of the Revolution, and of the restoration to peace. The power of asking leave of
absence, under such an accumulation of duties
as M. Cuvier's, was so rare, his time was so precious to himself, and the assurances of perfect
tranquillity in Paris, combined with the safety of
those whom they loved, were so decided, that
he and Mademoiselle Duvaucel determined on
proceeding to England. Instead, however, of
making a stay of six weeks, as they had at first
intended, they returned in a fortnight; and to
the happiness of those around him, M. Cuvier
found himself, even under the government of
the citizen king, in possession of all his honours,
his dignities, and his important functions.

In 1832 Baron Cuvier was made, by order of
Louis-Philippe, a peer of France, and the appointment of President to the entire Council of
State only waited for the royal signature, when,
on the 13th of May, of the same date, the noble
being closed his earthly career.

PART II.

THAT portion of my work which now lies before me has a grandeur and extent of subject which none but the life of M. Cuvier could present, and though I have confined myself to a mere description of his scientific labours, it will, in size, exceed all the others. But thus to follow him through this part of his vast career, thus to show him in the light of a savant, is no easy task; for, though a simple catalogue of his publications might have astonished by its length, it would have been very inadequate to my purpose. I have therefore attempted to carry my readers through each undertaking, by giving the outline of every plan, its purport, and its mode of execution; citing M. Cuvier's own sentiments and reflections in order to confirm that which is set forth, and occasionally giving even his own words, as examples of that style which was part of himself. I have also deemed it advisable to point out, in as brief a manner as possible, the state of natural history at the time he appeared, that a better estimate may be formed of the important revolutions which he either completed, or for which he laid the foundation.

Notwithstanding the great endeavours made in the earlier part of the seventeenth century towards the progress of natural history, as a science, there yet remained, when M. Cuvier first entered the learned world, as much to be done as had been effected since the revival of letters. The perfect form in which plants can be preserved with comparatively little trouble, the small expense at which they can be procured, and the narrow compass in which collections can be contained, gave them great advantages over other branches of natural history. Accordingly, we find that Botany had most profited by the exertions of several illustrious naturalists; it had even assumed that grouping, according to general organisation and structure, which is called the natural system; but Zoology, from the greater difficulties which the study of it presents, was, comparatively speaking, in a much less advanced state. On looking back to the history of this science from the beginning, we shall see three great names, the possessors of which caused the most important revolutions, who gave fresh impulse towards its perfection, and who have been the oracles of the civilised

E

world. To be able to mark the differences of one being from another is the foundation of this science; the great number of these beings necessitates classification, in order to assist the memory, and facilitate a perfect comprehension of their nature and properties, and the part they perform in creation. To Aristotle belongs the honour of the first epoch, by having invented the true method, that alone which can be permanent, as it is founded upon organisation, and is the result of personal observation. The writers after him, till the northern barbarians for a time buried all letters in obscurity, contented themselves with copying what he had done from one work into another, and by no means followed his example of seeing and judging for themselves. During the middle ages, now and then an enlightened monk, for a moment, threw a glimmering light over some branch of animated nature, and the first revival of learning presents us with many able efforts in this department of science. At length Linnæus appeared, and formed the second era. He assembled all known living beings together, and classed them according to the mass which he thus brought before him, selecting one or two individual characters as the foundation of his

clear and simple system, and by this, and by his
ingenious binary nomenclature, not only accomplished the great object of natural history, which
is to make us acquainted with the beings themselves, but by thus collecting them together,
greatly contributed to our knowledge of their
affinities. It was easy to be seen, however, that
in proportion as our knowledge of nature increased, this artificial classification would scatter
so many groups that were intended to remain
united among themselves, that it would be
found insufficient for the enlarged scale which
the discoveries of every year presented to us.
The Systema Naturæ then of Linnæus became
a mere sketch of what was to be done afterwards; even more recent naturalists touched
with a timid hand upon the natural grouping of
the highest branches of the science, and it was
reserved for a mighty genius of our own time to
open the path to us, and to smooth the difficulties of that path, by precisely determining
the limits of the great divisions, by exactly
defining the lesser groups, by placing them all
according to the invariable characters of their
internal structure, and by ridding them of the
accumulations of synonymes and absurdities

which ignorance, want of method, or fertility of imagination had heaped upon them.

Gifted with natural powers beyond the common lot of mortality, guided in earliest youth by a sensible and rightly judging parent, and prepared by an excellent German education, M. Cuvier was still further aided by a circumstance which, at first sight, seemed to be an obstacle to his progress. Almost excluded from the society of first-rate naturalists, and deprived, by the distracted state of France, of access to first-rate books, he was driven to nature herself; and as she, in her most minute operations, carries into execution that beautiful order and perfection which distinguishes her larger productions, so, to talents like those of M. Cuvier, did the study of the most insignificant animals open a vast field for future research and investigation. His mind was peculiarly calculated to embrace the great whole which a mass of details offers; at the same time he knew, that by an intimate and accurate knowledge of these details alone could he realise the comprehensive views which, even in his first studies, filled his great mind. He was of opinion, that every branch of science was to be rendered important if studied properly; no one, therefore, set a

higher value on minutiæ, at the same time he was
never once seen to lose himself in the intricacies
and minor considerations attached to these minutiæ.   Every research, no matter how humble,
how insignificant it might appear to the eyes of
others, was by him converted to the furtherance
of his great objects, the discovery and just appreciation of the truth.

The anatomical labours of M. Cuvier tended
to determine the physical functions of every
animal, of each part of each animal, and to as-
sign to the animal itself its place in the series of
beings ; to prove, that as each of the parts of
an organised being has a function to perform, so
does each being play its part in nature, acting
on all that surrounds it, and contributing to
form that whole in our planet, which excites the
wonder and admiration of all enquirers ; a whole
which, perhaps, takes its station in the parts of
a still wider expanse, into which we cannot penetrate.   " All is linked together," said M.
Cuvier, speaking of creation, " all is dependent,
all existence is chained to other existence, and
that chain which connects them, and of which
we can only see some comparatively insignificant
portions, is infinite in extent, space, and time."
He believed that all things in this world were

made for some express purpose; he believed
that all was due to one Supreme Intelligence,
which had provided organs for fulfilling the
ends for which all things were created. His
method resembled that of Aristotle, Bacon, and
Newton, for it was that of observation and experience, and, like them, he felt that no general
formula could be founded, no general principle
could be established, without a vast assemblage
of facts. He not only rejected all theories which
were not thus founded, from a conviction that
they led the mind astray from real observation,
but he carefully abstained from encouraging
any system which resulted from the discovery
of only a small number of facts; believing that
systems so based led their followers solely to
study those facts which were favourable to their
own peculiar views.

These were the broad principles which M.
Cuvier applied to every branch of human knowledge; for, like the Greek philosopher, he was
not ignorant of any thing, not even excepting
the mathematical sciences, of which he understood the foundation and machinery as if he
had studied them in the character of a professor. That same intelligence, also, which comprehended the form and organisation of the

beings of the present and former world, had
penetrated into the organisation of political
bodies, and perfectly appreciated their springs
of action, their strength, and their weaknesses.
Thus gifted, thus instructed, M. Cuvier unconsciously became a central point, round which
the scientific and learned of every class sooner
or later rallied. He was the kind and equitable
oracle of savans of all countries; for, wholly divested of national prejudices, and delighting to
dwell on that which was noble in all mankind,
he was never, for an instant, obscured by party
spirit, and was wholly unconscious of that su-
percilious feeling of superiority, which is so
hurtful to the progress of its possessor, and also
to the progress of others.

The earliest of M. Cuvier's scientific labours
were directed towards Entomology, and in them
we behold the dawning efforts of his genius,
the foundation of that minute and detailed observation which so particularly distinguished all
his researches, and of which I am about to give
rather a lengthened description, in order to
show that he commenced the task before him
in a way that necessarily led to the perfection
he afterwards attained. He has been heard
to observe, that the wonderful things he met

with in the organisation of insects raised his
genius to elevated thoughts; and such was his
opinion of Entomology in later life, that he asserted, " If I had not studied insects from
choice when I was at college, I should have
done so later, from a conviction of its necessity."
An anecdote is related of him by M. Audouin*,
in his Discourse, read at the Entomological Society of Paris, which proves still further the
value he set upon such pursuits. A young
student of medicine came to him one day, and
ventured to tell him, that he had discovered
something new and remarkable in dissecting a
human subject. " Are you an Entomologist?"
asked M. Cuvier.—" No," replied the student.
—" Well, then," returned M. Cuvier, " go and
anatomise an insect, I care not which, the
largest you can find, then re-consider your observation, and if it appear to be correct, I will
believe you on your word." The young man
submitted cheerfully to the proof; and soon
after, having acquired more skill and more
judgment, went again to M. Cuvier, to thank
him for his advice, and, at the same time, to

* Professor of Entomology at the Jardin des Plantes,
having succeeded to the chair recently vacated by the death
of M. Latreille.

confess his error. " You see," said M. Cuvier, smiling, " that my touchstone was a good one."

In another part of this work I shall have occasion to speak of the Entomological drawings of M. Cuvier, but this is the place to show to what extent he carried these youthful researches. Several fragments and memoirs, from his pen, exist on this subject; and among them is a paper, written in Latin, at the age of twenty-one, while in the château de Fiquainville, describing several Carabi*, and accompanied by illustrations, which were executed with the utmost delicacy and fidelity. Several magnified details were added to the text, which were prior to many afterwards given as new by professed Entomologists. In the same paper were delineations of other Coleoptera, and also of several Hemiptera, and various insects, accompanied by descriptions. In 1791 M. Cuvier corresponded on the same subject with MM. Fabricius and Pfaff; and wrote various papers concerning Pediculi and other parasitical insects. Some drawings, probably made about this period, were afterwards given by M. Cuvier

* A tribe of insects which takes its place in the great order, most commonly known under the name of Beetles.

to M. Lamarck, consisting of the most beautiful representations of Crustacea, forming twen-
ty-three separate pages, and containing, among
native marine Crustacea, several exotic species.

On coming to Paris, one of the first works
communicated to his friends by M. Cuvier was
a memoir, on the formation and use of a method
in pursuing the study of natural history, and
which he applied most happily to insects. This
memoir was followed by several more especial
labours, among which may be remarked, the description of a species of wasp (*Vespa nidulans*),
originally from Cayenne. In this paper he corrected an error made by Reaumur, who described
and figured the Chalcis, a parasitical insect,
living in wasps nests, as the female of the *Vespa
nidulans*. Soon after the appearance of the
above, a very interesting memoir was published
on the Cloportes (*Oniscus*, Lin.) in which some
parts of the mouths of Crustacea were described
for the first time. This was soon followed by
several others; one of the most remarkable of
which was a critical dissertation on the species
of crabs known to the ancients, and on the
names then given to them. In the month of September, 1797, M. Cuvier read, before the Institute, a very curious dissertation on the manner

in which insects are nourished. Having established that the dorsal vessel is not a true heart,
and that it does not furnish any means of circulation, it was necessary to account for the way
in which the nourishing fluid is carried to all the
organs.   M. Cuvier proved that this juice passes
through the cells of the intestinal canal, that it
spreads over the interior of the body, and, en-
circling all parts, is secreted by simple imbibition.   In this memoir he also stated, that the
secreting organs of insects are not solid glands,
as in all those animals which possess a heart and
blood vessels, but that they are composed of
spongy tubes, sometimes folded back upon themselves, intimately united by tracheæ, and which
may be always unrolled when time and patience
are called in to aid the task.   All these observations were attended with a result which is
always gratifying in natural history; they
established insects in a very natural and distinct
class, and, like other well directed labours, and
well founded remarks, these discoveries induced
others to make the same researches, and a new
field was open to the Entomologist.   If M. Cuvier was at any time doubtful, he did not hesitate
saying so : he corrected himself when he had
been mistaken ; and even at this period, when

he had all his fame to make, so far from being annoyed at the endeavours of others, he was the first to encourage them, to give them his honourable suffrage, and to receive as friends those who ventured into his province, in order to settle a doubtful point of science.

The mode of circulation in the Annelides was not better determined than that of insects, and M. Cuvier also turned his attention towards them. It was in pursuing this enquiry that he told anatomists, that the red colour of the liquid contained in leeches does not in the least proceed from the blood which the animal has imbibed, but that it is their own blood which circulates in four principal vessels. This important observation separated leeches, and animals analogous to them, from those with white blood; and caused Lamarck to give the class to which they belong the distinct name of Annelides. In M. Cuvier's great work on Comparative Anatomy, all the peculiarities belonging to insects, and other articulated animals, were afterwards given; and as he carried his labours into a wider expanse, he left their external forms and classification to others, and confined himself solely to their internal structure.

After thus noticing the earliest scientific la-

bours of M. Cuvier, which, in fact, were the
preparations for all that followed, I think it best
to proceed to that on which he based the great
works of a later period, considering the Tableau
Elémentaire, and the two editions of the Règne
Animal, as different stages of the same work,
and, with the Fossil Remains, and Natural History
of Fishes, as the results of his discoveries in
comparative anatomy. The collection of M.
Cuvier's lectures on this subject is preceded by
an introductory letter, addressed to M. Mertrud, in which the author submits the plan of
his work, the necessity of such an undertaking,
acknowledges the assistance afforded to him, and
states the care with which he has revised the
whole, previous to its publication.

The first lecture is a sort of preliminary discourse, and bears the general name of Animal
Economy. It is, however, divided into five
heads, viz. Organic Functions, Structure of the
Organs, Differences of Organs, Affinities of
Organs, and Division of Animals. From this
first lecture I shall make a few extracts, which
may enable my readers to form some judgment
of the work.

After examining the nature of the principles
of life, the learned author establishes the general

conclusion, " that no body exists which has not
once formed part of a body similar to itself,
from which it has been detached; or, that all
bodies have shared the life of another body, before they themselves exercise vital motion; and
it is even by the effect of the vital force, to
which they then belonged, that they have become sufficiently developed to support an isolated
life." From this conclusion may be deduced
the axiom, " that life springs from life, and no
other life exists than that which has been transmitted from one living body to another, in
uninterrupted succession." . . . . . " Being
unable to go back to the first origin of living
bodies, we have no resource," says M. Cuvier,
" but to seek information concerning the true
nature of the forces which animate them, in an
examination of their composition; that is to say,
of their substance, and the combination of elements which composes this substance or tissue.
For although this tissue, and this combination,
are in some measure the results of the action
of the vital principles which gave them being,
and continue to preserve them, it is evident that
these principles can only have in them their
source and their foundation. Thus, if the first
assemblage of these mechanical and chemical

elements of a living body has been effected by
the vital principle of the body from which it descends, we cannot but find in it a similar force,
and the causes of this force, in order to exercise
a similar action in favour of the body, which, in
its turn, descends from it. But, although our
knowledge of the composition of living bodies is
too imperfect to deduce clearly from it the effects they present to us, we may, at any rate,
make use of that which we do know, in order to
recognise these bodies, even when inactive, and
to distinguish their remains after death; for in
no unorganised bodies do we find fibrous or cellular tissue, or that multiplicity of volatile elements which forms the characters of organisation,
whether actually living, or having lived. Thus,
while inanimate solids are only composed of po-
lyhedral particles, mutually attracted by the
faces they present; while they only resolve themselves into a limited number of elementary substances; while they are only formed by a combination of these substances, and an aggregation
of these particles; while they only increase by
the juxta-position of new particles, which envelope the first mass by their layers; and while
they are only destroyed by some mechanical or
chemical agency, which alters their combin-

ations; on the other hand, organised bodies,
composed of a tissue of fibres and plates, the in-
tervals of which are filled with fluids, resolve
themselves almost entirely into volatile substances, spring from bodies similar to themselves,
from which they are only separated when they
can act by their own strength, assimilate themselves incessantly with foreign substances, and,
introducing these substances between their particles, increase by internal force, and at length
perish by this internal force, by the effects
even of their vital principle.   To originate in
generation, to increase by nutrition, and to end
by death, are the general and common characters of all organised bodies; but if several of
these bodies only exercise these and their necessary functions, and have only the organs requisite for this comparatively limited part in creation, there are many others which exercise
peculiar functions, which not only require organs
particularly adapted to them, but induce a mo-
dification in the general functions.   Of all these
peculiar functions, feeling and moving at will
are the most remarkable, and most influence the
other functions.   Independent of the chain
which links these two faculties, and the double
set of organs which they require, they yet carry

with them several modifications into the functions common to all organised beings, and these
modifications more particularly belong to and
constitute the nature of animals."

As one example, among many others which
the limits of this volume will not allow me to
insert, I shall cite M. Cuvier's general description of digestion.   " Vegetables, which are attached to the ground, absorb the nutritive parts
of the fluids which they imbibe by means of
their roots.   These roots, divided to infinity,
penetrate into the smallest spaces, and, as it
were, seek at a distance for nourishment to the
plant to which they belong: their action is tranquil and continuous, and is only interrupted by
a drying-up of the juices in the soil which are
necessary to them.   Animals, on the contrary,
not being fixed, and constantly changing place,
must carry with them the provision of juices
essential for their nutrition; therefore they have
received a cavity in which their alimentary substances are placed, into the cells of which open
the pores, or absorbing vessels, and which, according to the forcible expression of Boerhaave,   are
true internal roots.   The size of this cavity, and
its orifices, permit several animals to introduce
solid substances into it; these require mechan-

F

ism to divide them—liquids to dissolve them;
and nutrition no longer commences by the immediate absorption of substances as they are
supplied by the ground and the atmosphere; it
must be preceded by a multitude of preparatory
operations, the whole of which constitute digestion."

From the second division of this first lecture,
which treats of the organs of which animals
are composed, I shall select the passage concerning the senses, as most interesting to the
general reader. After exposing the nervous
system in its different bearings; after noticing
the cellular tissue, the medullary substance, the
muscles, the bones, the joints, the chemical ana-
lysis of various parts of the body, &c., M. Cuvier proceeds:—" We perceive the action of external bodies on our own, in proportion as the
nerve which is affected by them communicates
with the spinal chord, or common bundle of
nerves, and this with the brain; a ligature, or a
rupture, by intercepting the physical communi-
cation, destroys the feeling. The only sense
which belongs to all animals, and which exercises its influence over nearly the whole of the
surface of the body of each, is the touch. It
resides in the extremities of the nerves which

are distributed through the skin, and makes
known to us the resistance of bodies and their
temperature. The other senses seem to be but
more elevated modifications of the touch, and
are susceptible of more delicate impressions.
Every one knows that they are the sight,
which resides in the eye; the hearing, which
resides in the ear; the smell, which resides in
the membranes inside the nose; and the taste,
the seat of which is in the teguments of the
tongue. These are all situated at the same ex-
tremity of the body which contains the brain,
and which we call the head, or chief. Light,
vibrations of the air, volatile emanations floating
in the atmosphere, and saline particles soluble
in water, or the saliva, are the substances which
act on these four senses, and the organs which
transmit their action to the nerves are especially
adapted to each. The eye presents transparent
lenses to the light, which break its rays; the
ear offers membranes and fluids to the air, which
receive its agitations; the nose draws up the air
which goes to the lungs, and in its passage
attracts the odorous vapours contained in it;
and, lastly, the tongue is furnished with spongy
papillæ, which imbibe the savoury liquids offered to it. It is by these means that we are

conscious of the things and circumstances which
pass around us, and of the vast number of those
which pass within us; and, independent of the
internal pains which warn us of some disorder
in our organisation, and the sufferings occasioned
by hunger, thirst, and fatigue, it is by means
of these senses that we feel the emotions of pity,
the agonies of fear, &c. These latter sensations
are rather the effect of a reaction on the nervous
system than immediate impressions; and as the
sight of some imminent danger makes us fly
without the will having had time to act, it is
also involuntarily that we feel transport at the
sight of a beloved object, or shed tears at the
sight of virtue in distress. These effects of the
nervous system arise from the numerous communications of particular nerves, called sympa-
thetic, existing between divers ramifications of
the general trunk; and by means of which impressions are more rapidly transmitted than by
means of the brain. These knots of nerves,
which, when enlarged, bear the name of ganglions, are a species of secondary brains, and are
always of greater size, and in a greater number,
as the proportion of the principal brain is less
considerable."

When, in the third division, M. Cuvier treats

of the differences of the organs of animals, he
observes, that the circulation of the blood furnishes the most important variations. " First,
there are animals which have no blood, such as
insects and zoophytes; and, secondly, those
which have it, possess it in a double or simple
mode. That is called double circulation when
no part of the venous blood can re-enter the
arterial trunk until it has made a certain circuit
in the organ of respiration, which must be
formed by the expansion of two vessels, the one
arterial and the other venous, nearly of equal
size, but shorter than the two principal vessels
of the body. Such is the circulation of man,
mammalia, birds, fishes, and many mollusca.
In simple circulation, a great part of the venous
blood re-enters the arteries without passing
through the lungs, because this organ receives
but one expansion from one branch of the arterial trunk. Such is the circulation of reptiles.
There are yet other differences in the existence
and position of hearts, or muscular organs destined to impel the blood. In simple circulation there is never more than one; but when
the circulation is double, one part is sometimes
seated at the base of the principal artery, and
the other at that of the pulmonary artery; and

sometimes there is only one of these two parts. In the first case, the two hearts, or, rather, the two ventricles, may be united in one single mass, as in man, mammalia, and birds; or they may be separated, as in the cuttle-fish. When there is only one ventricle, it is either placed at the base of the artery of the body, as in snails and other mollusca, or at the base of the pulmonary artery, as in fishes.

" The organs of respiration are equally fertile in remarkable differences. When the element which is to act on the blood is the atmospheric air, it penetrates even into the interior of the respiratory organ; but, when it is water, it simply glides over a surface, more or less multiplied. These surfaces, or leaflets, are called branchiæ, and are found in fishes, and many of the mollusca. Instead of leaflets, there are sometimes tufts, or fringes. Air penetrates into the body by one or several orifices. In the first instance, which is that of all animals with lungs, properly so called, the channel which receives the air is subdivided into a multitude of branches, which terminate in as many little cells, generally collected into two masses, which the animal has the power of compressing or dilating. When there are several openings, which we see only in

insects, the vessels which receive the air are
ramified to infinity, in order to carry it to every
part of the body without exception, and this is
what is called respiration by means of tracheæ.
Lastly, the zoophytes, with the exception of
echinodermes, have no apparent organ of respiration."

In the third portion of this opening lecture,
the affinities of organs are described, and their
manner of acting on each other. " Of what
use," says M. Cuvier, " would sensation be to
us, if muscular force did not help it, even in the
most trifling circumstances? What use could
we make of touch, if we could not carry our
hands towards the palpable object? and what
should we behold if we could not turn our eyes
or head at pleasure? It is on this mutual de-
pendence between the functions, and on this
reciprocal aid, that the laws are founded which
determine the affinities of the organs of animals;
which affinities are as necessary to them, as
metaphysical or mathematical laws are to other
parts of the creation. For it is evident, that a
suitable harmony between those ,organs which
act upon each other, is a necessary condition to
the existence of the being to which they belong;
and that if one of these functions were modified

in a manner incompatible with the modifications
of others, that being could not exist.   Modern
experiments have shown, that one of the principal uses of respiration is to reanimate muscular
force, by restoring to the muscular fibres their
exhausted irritability, and, in fact, among the
animals which breathe the air in a direct manner, we see those with double circulation, and
not an atom of whose blood can return to the
parts till after it has been respired.   Mammalia
and birds not only live always in the air, and
move in it with more force than other animals
with red blood, but each of these classes enjoys
the faculty of moving, precisely according to
the quantity of its respiration. Birds, for instance, are as much impregnated with air within
as without; not only the cellular parts of their
lungs are very considerable, but these organs
have bags, or appendices, which are prolonged
throughout the body.   Thus, in a given time,
they consume a much greater quantity of air in
proportion to their size than quadrupeds; and
doubtless it is this which gives to their fibres a
prodigious and instantaneous force, and which
renders their flesh fit to act powerfully on those
violent movements which sustain them in the
air, by the simple vibrations of their wings."

In the concluding part of this first lecture,

treating of the Division of Animals, M. Cuvier
sums up the great characters of the animal kingdom, proceeding from Mammalia to Zoophytes,
or,. in other terms, the whole range of animal
life, from man, to the simple pulp which scarcely
merits the name of an organised being. From
this I do not find it possible to cite any isolated
passage, the whole is so beautifully linked together; but the perusal of this portion alone is
calculated to enlarge our ideas respecting creation, to give us new thoughts concerning the
common occurrences of life, and to lead us to a
train of reflections, which mount upwards to the
great Source of that life which is presented to
us in so many extraordinary and elaborate
forms. The rest of the work consists of a description of these forms, external and internal;
the minutest details concerning the use of each
organ are also given to us; the chemical composition of each part is explained; the greater or
lesser developement of this wonderful machinery
and combination is set forth; the total absence
of some parts; the results of these differences,
and the action of the whole in the vast field of
nature, all are laid before us with a clearness
and precision that are truly admirable. For, although endowed with imagination, brilliancy of

ideas, and eloquence of language, M. Cuvier
has in this, as well as his other scientific labours,
affirmed only "that which he has seen and
touched," and, far from wishing to persuade by
other means than positive evidence, he presents
his readers with nothing that can draw the mind
from the contemplation of reality. From this
work we may deduce certain general rules, certain axioms, which may apply to every part of
animal life, in every corner of the world. Let
us take the single instance of a person ship-
wrecked in an unknown sea, and cast up by the
waves on a shore wholly strange to him. Towards the means of life are directed the efforts
of returning consciousness: vegetables will first
offer themselves to his notice, as most easily
procured; but an anatomist will know, that his
teeth and organs of digestion were given to him
that he might repair the exhaustion of his frame
by animal substances, and that without these he
will not be so healthy and strong as nature
intended he should be. A single bone, accidentally lying in his path, will tell him, if this
comparatively desert country contains animals
against which he must provide means of defence; but what will be his joy, if this single
bone informs him, that ruminating animals exist

there.   Milk, flesh, beasts of burden, skins for
bedding and clothing, at once present themselves to his view.   Furnished with such sources
of comfort, he is prepared to avoid the destructive, to ensnare the swift, and to make use of the
docile;   and weaker in bodily force, perhaps,
than the animals by which he is surrounded in
his desolate habitation, yet, by the superiority
of his intelligence, he becomes their sovereign.

To say precisely what this great treatise dis-
plays, in an extent of five thick octavo volumes,
each containing from five to six hundred pages;
to give an exact list of every thing it embraces,
would be to offer a dry catalogue of names,
which would not be generally understood; but in
order to show the manner in which it is conducted throughout, and how thoroughly it
carries the student into every portion of an
organised body, I submit the way in which the
head is treated.   The different bones which
form the box called the skull, with their shapes,
are first detailed; then follow the articulation of
the head upon the spine, and its consequent movements; the muscles which aid these movements,
and give force and motion to the jaws; the un-
equal surfaces of the interior of the skull; the
holes of the skull; the bones of the face; the

brain and its coverings; the nerves which proceed from it; the different parts of the eye, and its consequent vision; the muscles which move the eye; the eyelids; the ear, and its complicated parts; the muscles and nerves belonging to it; the movements of the lower jaw; the teeth in all their stages; the salivary glands; the bones of the tongue, its muscles, and the tongue itself, &c. What a task thus to carry us through all creation! And yet the noble author considered this merely as a foundation for one of still greater magnitude, the preparations for which he had been carrying on during the whole of his active life, and the results of which preparations were shortly to have been given to us, had not death suddenly arrested his progress — had not the inscrutable will of the Almighty suddenly closed upon us the way which this great genius had opened to our progress.

To the above work was awarded one of the " prix décennaux," instituted by Napoleon in 1810, an account of which may be acceptable to the English reader. Wishing at that time to divert the public attention from passing events (the Spanish campaign, &c.), the Emperor issued a decree, stating, that as he was desirous of rewarding and encouraging every species of study

and labour, which could contribute to the glory of
his empire, he had resolved to bestow prizes of
money, every ten years, on the 9th of November,
on the best works in every branch of science, art,
and literature. The prizes were to be proclaimed
by the Minister of the Interior, and the successful candidates were also to receive a medal from
the hands of the Emperor himself, in presence
of the princes, the dignitaries of the state, the
great officers of the University, and the whole
body of the Institute, assembled at the Thuilleries.
All labours having sufficient merit were to be
examined by a jury and judges, composed of the
presidents and perpetual secretaries of the four
classes of the Institute.   Each class to make a
catalogue raisonné of the works put to the suffrage ; those deemed worthy of approaching
the prizes, to receive honourable mention ; but
those of sufficient merit, in the opinion of the
judges, to obtain the prize, to be noticed with
still greater detail. All the reports and discussions to be given to the Minister of the
Interior, by whom they were to be kept entirely secret from the public. No judge to be
allowed to pronounce on the merits of his own
productions. These prizes soon became an universal theme; an exhibition of the pictures

painted for them took place in the Louvre, and every body was more or less interested. The juries sat, the judges pronounced sentence; and because the Comparative Anatomy proceeded from one of the latter, though it received the praise due to it, the prize for this subject was awarded to another work. Delay took place, and the Emperor deemed a revision of the judgment necessary. During this revision M. Cuvier was in Italy, and advantage was taken of his absence to change the sentence, and recommend the prize to be bestowed on him. The greatest freedom was given to discussion, in the idea that all would be strictly confidential on the part of the government; when, to the astonishment of every one, the whole of the reports given to the Minister of the Interior was published in the Moniteur. Could any thing be better calculated to accomplish the desires of his Imperial Majesty? No sooner did the affair languish, and people cease to talk of it, from the conviction that all was done, than he set the whole capital in a turmoil of bickering and dispute; for every one had either his own cause, or that of his protégé to defend. The result proved it to be one of those master-strokes of policy of which Napoleon was so capable; and what was his intention throughout is very evident, for the prizes were never even mentioned afterwards. The reports, however, have been collected, and form a very curious quarto volume.

From the writings on Comparative Anatomy, I naturally turn to that vast collection of the subjects themselves, formed by M. Cuvier at the Jardin des Plantes; and when I repeat, that this collection was not only the principal source from which he drew the materials for the great work just mentioned, but was the basis for most others, it is scarcely necessary for me to enter into many details concerning it : to its leading features I shall therefore confine myself. It is contained in fifteen rooms of various sizes ; and in these fifteen rooms we may verify almost every fact stated by M. Cuvier, by actual inspection ; and we are lost in admiration, not only at the vast operations of nature, but at the mind which appreciated them, and made them known to his fellow men. The collection should be viewed by beginning at the room up stairs, which is farthest from the entrance, and which communicates with M. Cuvier's house. In this are the Mollusca, and at once assuming the character of a person wholly ignorant of anatomy, I cannot do better than describe the probable impressions

of this person, as he follows the suite of rooms.
His astonishment will be first excited by finding,
that such mis-shapen masses as the common
oyster possess liver, heart, lungs, &c.; he will
wonder at the various peculiarities presented by
the inhabitants of the shells he has been accustomed to find on the sea-shore, and to consider as mere toys; he will be surprised at the
number of those insects which exist only on
living bodies, and all disgust will be lost, in contemplating the variety of their forms. The two
next rooms will present to him that complicated
machinery which is contained in beings of a
higher order, by which they reanimate their
strength; by which, in fact, they live. A step
farther, and he will see the muscles fortified and
brought into action by that very machinery
which he has been examining. But the organs
of the senses will have previously arrested his
attention; for he there beholds in the eye the
very powers he is exercising, and which are affording him such infinite gratification. The ear,
which gives so much pleasure, and frequently so
much pain; the voice *, by which we impart our

* After reading a very interesting Memoir on the organs
of the voice in birds, before the Institute, in 1798; a very

own feelings; the reason why the power of
uttering those delicious tones which captivate
and soothe us into harmony, those impassioned
sounds which cheat us into an entire forgetfulness of aught but themselves; those accents of
fury which frighten us to agony, or those grave
and calm communications of the mind, are only
given to man; are all there, and wonder succeeds
to wonder, leaving it difficult for the stranger to
decide in which room he finds most interest.
That part of the human frame from which we
suffer most, the teeth, and dentition, in all its
stages, and in all animals gifted with it, are laid
open to his view, with the important characters
they afford for classification, and the progress
made from the concealment of the tooth in its
socket at the birth of the infant, to the filling up
of the empty sockets into one solid mass, in the
aged person. Close to human teeth are the

---

celebrated anatomist present exclaimed, that M. Cuvier had
been wrong in stating, that physiologists had not yet agreed
concerning the mechanism of the human voice, which some
compared to a wind, and others to a stringed instrument;
for that this question was now decided in favour of the wind
instrument. " You are deceived," involuntarily cried another
equally learned anatomist; " it is a stringed instrument."
This second observation caused a general smile, for it proved,
most unexpectedly, the truth of M. Cuvier's assertion.

G

enormous and solitary grinders of the two living
species of elephants, the unchanging ivory of
the tusks of the walrus, the fearful weapons of
the lion and the tiger, and the sharp incisors of
the bat. How surprised will the novice be, to
find, that the head, which he has been accustomed to consider as one mass of bone, is, in
mammalia, composed of several parts, and in
fishes, divided to infinity. Deeply will he reflect,
when, in an adjoining room, he follows, in the entire skull, the gradations of the frontal bone,
which mark the most intelligent of mankind, to
the animal whose only instinct is that of procuring food; and on descending the staircase, to
find himself in the midst of human skeletons, in
all their varieties, from the Hottentot Venus to
the graceful being of an European drawing-room.
For a moment, his feeling of admiration at the
works of God, are interrupted by a contempt of
that external beauty which has hitherto been so
precious in his eyes; but the great room, if I
mistake not, will banish every sentiment but
those of awe and reverence; for he will there
find himself walking amid the remains of the
most gigantic of the earth, and the enormous
monsters which inhabit the depths of the ocean.
The solid and ponderous members of the ele-

phant, the long neck of the giraffe, the massive bulk of the whale, the hand-like fins of the dolphin, the strength and vigour of the horse, the solemn force of the bull, and the light and elegant action of the antelope, may all be traced in these frame works of creation ; and as the visitor quits the galleries, I think I cannot be wrong in supposing, that he will own his personal insignificance in the great scale, his conviction of the adaptation of nature to all the purposes for which it is intended, and will learn to respect that being of his own species, who, by his influence, his personal exertion, amassed, and, by his wisdom, set before him, the marvellous works which he has just been contemplating.

With so perfect a knowledge of the formation of living beings, it scarcely seems surprising that M. Cuvier should have made those deep researches among the fragments of the former inhabitants of the globe, by which his great name has been associated with every labour relative to the construction of the earth. For although the researches of De Saussure, De Luc, Pallas, and Werner, appeared to have brought geology to the highest perfection it could attain, it was M. Cuvier who gave the impulse, who made a science of fossil organic remains. His powerful comprehension, at the first glance, measured the extent of the science, appreciated its importance, and foresaw the light it would shed over the formation of our planet. Already, in 1796, he read a Memoir, at the Institute, which contained his suspicions, that no species of those fossil remains, so abundant in the northern parts of the world, belonged to animals now existing. He even then thought that they had formed beings which had been destroyed by some revolution of the globe, now replaced by others, perhaps equally to be destroyed. With a view of ascertaining the truth of these suppositions, he sought every means of determining the species, genera, and classes of these relics, by an unwearied inspection of all that could be found, by making himself acquainted with the discoveries previously made, by exactly ascertaining the localities where these remains had been dug up, the nature of the soils in which they had been enveloped; and he eloquently invited all the savans of Europe to aid him in his great enterprise, impressing on them the importance of these researches, and requesting them to report their labours to him, which labours he promised to state in his work, and which promise he faithfully performed. In the Prelimi-

nary Discourse to the Fossil Remains (which
has been published in a separate form, has undergone several editions, and been translated
into almost every modern language, under the
title of " Theory of the Earth"), treating of the
revolutions of the globe, he says, " Antiquary of
a new species, I have been obliged at once to
learn how to restore these monuments of past
times, and to decypher their meaning. I have
been obliged to collect and bring together the
fragments which compose them into their primitive order; to reconstruct these ancient beings; to reproduce them, with their proportions
and characters; and, lastly, to compare them
with those which now live on the surface of the
globe."

To this task did M. Cuvier devote a large
portion of his life, and his first care was, to determine the living and fossil species of elephants,
which form the subject of the first volume. The
plan he adopted was, to describe the osteology
of the best known species; to point out the
countries they inhabit; to ascertain how many
species have been found; and, then, to compare
them with those bones which are in a fossil state.
He himself visited many of the spots whence
these remains had been taken; such as Eng-

land, Holland, Germany, and Italy; and others
were brought to him, in order that he might be
an eye-witness of every thing which he endeavoured to prove.   These researches entirely set
at rest the question concerning the existence, or,
rather, the finding of human fossils.   Such relics have never, as yet, been discovered; and
the Guadaloupe skeletons, which have been so
much talked of, had probably been deposited
in that place after shipwreck; the soil by which
they were enveloped being of too recent a formation to admit of any idea that they were true
fossils, and the positions in which they laid, not
allowing of the supposition that they had been
purposely interred there.   Also, the pretended
histories of giants are, in this volume, entirely
refuted; and amusing accounts are there given
of the ignorance and credulity which caused
them to be so generally circulated; but on this
occasion, as, in fact, all others, M. Cuvier's own
words are the best, and he writes as follows:—
" The bones of elephants having more resemblance to those of man than they have to those
of other animals, even skilful anatomists have
been often tempted to take them for human
remains, and this probably occasioned the pretended discoveries of the tombs of giants, men-

tioned by ancient authors, and those of the middle ages." This example was unfailingly followed by more modern writers, for the marvellous is delicious food to the minds of most people. The great propagator of the *on dits* of natural history, Pliny, was not, of course, wanting on this occasion; and he speaks of the supposed body of Orestes as being thirteen feet three inches long. Few countries have been without these fables, and (to continue M. Cuvier's account) " one of the most celebrated was that of Teutobochus, in the reign of Louis XIII., which occasioned a number of famous disputes, in which the actors were much more anxious to abuse each other than to establish the truth. One of them, however, named Riolan, for a person who had never seen the skeleton of an elephant, showed, with considerable skill, that these bones probably belonged to such an animal. It would appear, as far as the fact can be now ascertained, that on the 11th of January, 1613, some bones were found in a sand pit, near the castle of Chaumont, or Langon, between the towns of Montricaut, Serre, and Antoine. Part of them were broken by the workmen; but a surgeon of Beaurepaire, named Mazurier, showed those which remained whole for money, in Paris and several other places, and, in order to excite further curiosity, he circulated a pamphlet, in which he asserted that they had been found in a sepulchre, thirty feet long, on which had been inscribed, 'Teutobochus Rex.' It is well known that this was the name of the King of the Cimbri who fought against Marius; and, to further this supposition, M. Mazurier added, that fifty medals were found in the same place, bearing the effigy of this Roman consul, and the initials of his name. The surgeon, however, was accused of having employed a jesuit, named Tournon, to write this pamphlet, and who forged the history of the sepulchre and the inscription. The pretended medals bore Gothic instead of Roman letters, and it seems that Mazurier never justified himself from these accusations of imposture." The bones were afterwards all recognised as belonging to elephants; but, notwithstanding this detection, there was no end to the stories about giants, and each country possessed its own marvellous tale. The city of Lucerne took for supporters to its coat of arms pretended giants found in 1577, near that place, and close by the cloister of Reyden, in a hole, which was accidentally formed by the tearing up of a large oak by the roots, in a

heavy gale of wind. The Council of Lucerne
sent them to Felix Plater, a physician at Bâle,
who had a drawing made of a human skeleton,
the size which he thought these bones indicated.
It measured nineteen feet, and was sent, with
the bones, back to Lucerne, where the drawing
is still preserved. It, and the bones still in existence, were recently inspected by M. Blumenbach, who recognised the latter as belonging to
an elephant.

But the champions of human fossils were not
contented with making them out of the bones
of elephants; and having found some animal
remains imbedded in slate, a few leagues from
the Lake of Constance, a learned physician
wrote a particular dissertation on them, entitled
" L'Homme Témoin du Deluge."—" It is not
to be refuted," said he, " here is the half, or
nearly the whole of the skeleton of a man, even
the substance of the bones, and, what is more,
the flesh, and parts still softer than the flesh,
are incorporated with the stone. In short, it is
one of the rarest relics we possess of that cursed
race which was buried under the waters." The
assertions of the learned Doctor, however, va-
nished before the penetrating eye of M. Cuvier,
who, judging from the relative form and propor-

tion of the bones, decided that this fossil was no
other than that of an aquatic salamander, of a
gigantic size and unknown species. In 1811,
having the power of examining the stone which
contained this " witness of the deluge," he, in
presence of several distinguished savans, and
with the drawing of a salamander before him, at
every stroke of the chisel verified his assertion.

But to return to the elephants : Asiatic Russia
swarms with these monstrous remains, and the inhabitants explain the phenomenon by supposing
that they belong to some living subterraneous animal partaking of the nature of the mole, and which
they call Mammout, or Mammouth. This fable
also extends to China. Besides the relics of
true elephants, found in America, there have
been yet two other gigantic animals discovered ;
the Mastodon and the Megatherium, the former
bearing great affinity to the elephant. These
animals have also formed a foundation for many
absurd stories, all of which have been refuted
by M. Cuvier's luminous researches : he states,
" that the great animal of Ohio was very similar
to the elephant in its tusks and its osteology,
with the exception of its jaws ; that it very probably had a trunk, but that in height it did not
exceed the elephant. It was, however, longer

than that quadruped, its limbs thicker, its belly
of less volume ; but, notwithstanding the little
importance of these differences, the peculiar
structure of its grinders suffices to establish it as
a separate genus. It was nourished nearly in
the same manner as the hippopotamus and wild
boar, but it did not occasionally live in the
water, like the former. It preferred roots, and
the fleshy parts of vegetables, which species of
food led it to seek an open or marshy country."
The bones of the Mastodon Angustidens are
much more common in North America than
elsewhere, and, perhaps, those of the great mastodon exclusively belong to that country. They
are better preserved and fresher than any other
known fossils, and, nevertheless, there is not the
least authentic testimony calculated to make us
believe, that there is still in America, or elsewhere, any living individual. Therefore, the
accounts published, from time to time, in the
American papers, concerning those that have
been met with wandering through the vast forests, or over the immense plains of this continent, have never been confirmed, and may be
consequently regarded as mere fables.

After having acquired vast experience in the
connection of organised beings with the soils in

which they have been preserved, and having
decidedly proved, that the more ancient the
formation, the more distant are its organic remains from those now existing, M. Cuvier determined to observe and describe all those contained in a limited circumference round Paris.
Already had he employed an intelligent workman*, whom he himself paid, in the quarries at
Montmartre, to collect the bones for him which
were almost daily found in that spot. He
spared no expense, rewarded all contributors
with the greatest liberality, and joyfully spent
considerable sums on that collection, which,
when his publications had given it the highest
value, he afterwards presented to the Museum
of the Jardin des Plantes, only receiving in
return, duplicates from the public library, of
those works which were wanting in his own
magnificent assemblage of books. Before M.
Cuvier found an opportunity of publishing his
discoveries, by means of the Annales du Muséum, and when the expense of employing professed artists would have been too much for his
means, he not only drew, but engraved the
plates himself; which precious proofs of his

* Named Varin.

talents are scattered through the work of which I am now speaking, but are more particularly contained in the third volume of the last edition.*

* Had I no motive of friendship and esteem to induce me to make known the merits of M. Laurillard, the secretary of M. Cuvier, it would be but justice to mention him here, as one who was associated with his patron in these and all succeeding labours, and who proved that the great anatomist carried his discrimination even into the mental organisation of humanity. The manner in which this association was formed is too interesting to be passed over in silence. M. Laurillard, also from Montbéliard, left his native place in order to cultivate his talents for design in the capital, with a view of becoming professional. He was there introduced to M. Fréderic Cuvier, for whom he executed some drawings. He also made one or two for M. Cuvier, without particularly attracting his notice. One day, however, M. Cuvier came to his brother to ask him to disengage a fossil from its surrounding mass, an office he had frequently performed. M. Laurilliard was the only person to be found on the spot, and to him M. Cuvier applied in the absence of his brother. Little aware of the value of the specimen confided to his care, he cheerfully set to work, and succeeded in getting the bone entire from its position. M. Cuvier, after a short time, returned for his treasure, and when he saw how perfect it was, his ecstasies became incontrollable ; he danced, he shook his hands, he uttered expressions of delight, till M. Laurillard, in his ignorance both of the importance of what he had done, and of the ardent character of M. Cuvier, thought he was mad. Taking however his fossil foot in one hand, and dragging M. Laurillard's arm with the other, he led him up stairs to present him to his wife and sister-in-law, saying, " I have got my foot, and M. Laurillard found it for me." It seems, that this skilful operation confirmed all M. Cuvier's previous conjectures concerning a foot, the existence and form of which he had already guessed, but

This edition consists of five quarto volumes, two
of which are divided into two parts; and among
the numerous lights thrown upon living objects,
and on the construction of the earth, we find
the resurrection of numerous species of mammalia, birds, reptiles, &c., making in all 168
vertebrated animals, which form 50 genera, and
of which fifteen are new. They have been
named by M. Cuvier, placed by him in the
range of created beings, and belong to every
order except Quadrumana, of which, as well as
the human race, not a single relic has yet been
found in a fossil state. All their localities have
been stated, and all the collections mentioned
where they have been preserved, with a laborious fidelity and extraordinary erudition. He

---

for which he had long and vainly sought. So occupied
had he been by it, that when he appeared to be particularly absent, his family were wont to accuse him of seeking
his fore foot. The next morning the able operator and
draftsman was engaged as secretary ; and M. Cuvier not
only attached to himself a powerful coadjutor, but an affectionate and faithful friend, devoted to him during life, and
now finding his greatest happiness in doing and saying that
which he thinks will most honour the memory of one so
loved and revered. He is appointed, by the will of M. Cuvier,
to finish and publish all the drawings they had made together
for the great work, which he called the " Grande Anatomie
comparée," — and most fervently must all followers of the
science wish for its appearance.

had, however, many difficulties to conquer,
among which was that of the incredulity of
others, who, being ignorant of the laws of organisation, of the necessary co-existence of
certain forms, did not comprehend how it was
possible to re-establish an animal from the fragments of its bones scattered through the layers
of the earth. How he triumphed will be gathered from the following extract from a letter
written to Dr. Duvernoy, a few days after a
meeting in which he had been obliged to dis-
cuss some particular objections addressed to
him. He thus wrote (1806), — " They have
just brought me the skeleton of an anoplotherium, which is almost entire, taken from Montmartre, and nearly five feet long. *All my conjectures have been verified,* and I find that the
animal had a tail, as long and as large as that of
a kangaroo, which completes its singularities."
For the furtherance of his inspection of the
neighbourhood of Paris, M. Cuvier associated
the learned geologist, M. Brongniart, with him
in his researches, who more particularly confined himself to fossil mollusca, and comparative
observations concerning other countries. The
principal geological result of these inspections
was to make known the fresh water deposits

above the chalk, each deposit covered by a marine deposit; irrefragable proofs of several irruptions and alternate retreats of the sea, in the
basin of Paris and its environs, since the period
of the chalk formation. This discovery was
solely due to M. Cuvier, and it was at Fontaine-
bleau that the truth suddenly flashed across his
mind. " Brongniart," he cried, " j'ai trouvé
le nœud de l'affaire." " Et quel est-il?" asked
M. Brongniart. " C'est qu'il y a des terrains
marins, et des terrains d'eau douce," replied M.
Cuvier.* It is most interesting to see how,
after many years of uninterrupted and difficult
investigation, of profound study and meditation,
M. Cuvier, in his beautiful Preliminary Discourse, sums up the facts which afford indisputable evidence of these great phenomena. " I
think," said the learned author, with MM. De
Luc and Dolomieu, " that if there be any
thing positive in geology, it is, that the surface
of our globe has been the victim of a great and
sudden revolution, the date of which cannot be
carried back further than from five to six thousand years; that this revolution has buried,

* "I have solved the difficulty."—" And what is it ? "—" It
is, that there are fresh water earths, and earths of salt water
formation."

and caused the disappearance of countries formerly inhabited by man, and animals which are
now known; and, on the other hand, has exposed the bottom of the water, and has formed
from that, the countries now inhabited . . . . but
these countries which are now dry had already
been inhabited, if not by man, at least by terres-
trial animals; consequently one preceding revolution at least must have covered them with
water, and, if we may judge by the different
orders of animals of which we find the remains,
they had perhaps been submitted to two or three
irruptions of the sea; and these irruptions, these
repeated retreats, have not all been slow or gradual. The greater number of the catastrophes
which brought them about have been sudden, a
fact easily proved by the last of all, the traces of
which are most manifest, and which has still
left in the North the bodies of large quadrupeds,
seized by the ice, and by it preserved, even to
our own times, with their skin, their fur, and
their flesh. Had they not been frozen as soon
as killed, putrefaction would have decomposed
them; and this eternal frost has only prevailed
over the places inhabited by them, in consequence of the same catastrophe which has de-

H

stroyed them : the cause, therefore, has been as
sudden as the effect it produced."

The ideas of M. Cuvier on the relative ages
of the strata of deposited soils, extending even
to different chains of mountains, have led to the
present system adopted by geologists, and from
them it may be concluded, that " all these
layers of deposited soils having been necessarily
formed in a horizontal position, the most ancient
are those which have been more or less raised
towards a vertical line by some catastrophe, and
the most recent are, on the contrary, the horizontal layers; because, having preserved their
original situation, it is evident that they could
only be formed after the revolution which
changed the position of those which are oblique,
which they more or less cover, and on which
they rest."

One of the most important questions treated
of in this work is that of the alteration in animal
forms ; whether the forms of lost animals, which
differ so much from those which are now living,
really indicate species and genera distinct from
species and genera now existing, or if time alone
has modified the primitive forms, so as to attain
the present form. The examination of this
question alone would give a satisfactory answer

(could they be convinced) to those who believe
in the indefinite alteration of forms in organised
beings, and who think that, with time and habits, each species might have made an exchange
with another, and thus have resulted from one
single species. However extraordinary and in-
comprehensible this system may appear to be,
which would take away the basis on which
science rests, and which could only be established by a definition of the possible duration of
a species in its original state, M. Cuvier seriously refutes it, and destroys it with one
objection, that of not finding intermediate modifications between an animal of the former and
present world, even when it approaches it most
nearly. He gives the definition of a species,
proves the constancy of certain conditions of
the forms which characterise it, and presents a
table of the variations which it is possible for it
to undergo. In short, he demonstrates, by a
scrupulous examination of the skeletons of mum-
mies, that the animals living in Egypt two or
three thousand years back, when compared with
those which now breathe on this classic ground,
have not, in the course of so many ages, undergone any important changes of form; that even
among the wild animals there has been no alter-

ation in the skeleton which could characterise
one race or variety. " There is nothing," to
use M. Cuvier's own words, " which can in the
least support the opinion, that the new genera
which I and other naturalists have discovered or
established among fossils, the Paleotherium, the
Anoplotherium, &c., have been the parent
stocks of some of the present animals, which
only differ from them in consequence of other
soil, climate," &c.   Further on he continues, —
" When I maintain that stony strata contain the
bones of several genera, and moveable earths
those of several species which no longer exist, I
do not pretend that a new creation has been
necessary to produce the existing species.   I
merely say that they did not exist in the places
where we now see them, and that they have
come from elsewhere.   For example, let us suppose that a great irruption of the sea shall now
cover the continent of New Holland with a
mass of sand, or other débris; the bodies of
kangaroos, wombats, dasyuri, perameles, flying
paalangistæ, echidnæ, and ornithorynchi, will
be buried under it, and it will entirely destroy
every species of these genera, since none of
them now exist in other countries.   Let. this
same revolution dry up the sea which covers

the numerous straits between New Holland and the continent of Asia: it will open a way for the elephant, the rhinoceros, the buffalo, the horse, the camel, the tiger, and all other Asiatic quadrupeds, who will people a country where they have been hitherto unknown. A naturalist afterwards living among them, and by chance searching into the depths of the soil on which this new nature lives, will find the remains of beings wholly different. That which New Holland would be in the above case, Europe, Siberia, and a great part of America are now, and, perhaps, when other countries, and New Holland itself, shall be examined, we shall find that they have all undergone similar revolutions. I could almost say, a mutual exchange of productions; for, carrying the supposition still further, after this transportation of Asiatic animals into New Holland, let us imagine a second revolution, which shall destroy Asia, their primitive country; those who afterwards see them in New Holland, their second country, will be as embarrassed to know whence they came, as we can be now to find the origin of our own."

I am aware that the extent of the work of which I am speaking can scarcely be recognised in the few extracts I am able to make, and it is

with a sort of fearfulness that I cite a few isolated passages, for fear of injuring the rest.
There must, however, necessarily be a degree of
imperfection where we can only judge by parts,
detached from a whole, which is so beautiful
when entire; and again impressing on my
readers that this volume is intended to lay before them the man himself, and describe his
labours, not to review or criticise them, I have
less hesitation in proceeding.

The gradual developement of great facts, the
doubts existing in the mind of the author at
certain periods of his progress, the confirmation
or dissipation of these doubts, the methods employed to ascertain the truth, the sacrifice of one
part of a fossil to verify another, the ingenious
contrivances for separating the remains from the
surrounding mass, the application of plaster models, which not only brought him faithful impressions of those which he could not procure,
from distant countries, but distributed his own
to every part of the world; are all related in
the course of the work with the most beautiful
simplicity.   When speaking of the sarigue*, M.
Cuvier says, " This rich collection of the bones
and skeletons of the animals of a former world

* A species of opossum.

is doubtless an enviable possession. It has been amassed by nature in the quarries which environ our city, as if reserved by her for the researches and instruction of the present age. Each day we discover some new relic; each day adds to our astonishment by demonstrating, more and more, that nothing which then peopled this part of the globe has been preserved on its present surface ; and these proofs will doubtless multiply in proportion as our interest in them is increased, and we are consequently induced to give them more of our attention. There is scarcely a block of gypsum, in certain strata, which does not contain bones. How many millions of these bones have been already destroyed in working these quarries for the purposes of building ! How many are destroyed by negligence, and how many escape the most attentive workman, from the minuteness of their size! Some idea of this may be formed from the piece I am going to describe. The lineaments there imprinted are so faint, that they must be narrowly examined in order to be recognised. Nevertheless, these lineaments are most precious, for they belong to an animal of which we find no other traces; to an animal which, perhaps, buried for ages, now reappears,

for the first time, to the eye of the naturalist."
At the end of the description of the sarigue,
M. Cuvier continues, — " I will not dilate on
the geological consequences of this Memoir*,
for it will be evident to all those who understand the systems relative to the theory of the
earth, that it overturns almost every thing which
concerns fossil remains.   It has been admitted
that the fossils of the North have been animals
from Asia; it was also allowed that the animals
of Asia had passed over into North America,
and had been there buried; but it appeared
that the American genera had come from their
own soil, and had never extended to the countries which now form the Old World.   My discoveries lead to the contrary opinion, and this is
the second proof I have received.   Fully per-
suaded of the futility of all these systems, I congratulate myself whenever a well-attested fact
destroys some one of them.   The greatest service that can be rendered to science is, carefully
and decidedly to find the place of every thing
before building upon it, then to begin by over-
throwing all those fantastic edifices which choke
up the avenues, and which prevent the entrance of

* It was first published as a separate Memoir in the Annales du Muséum.

those men to whom the exact sciences have given the excellent habit of relying solely on evidence, or, in a dearth of positive evidence, on circumstances, according to their degree of probability. With these precautions there is no science which may not almost become geometrical. Chemists have lately found this with regard to their pursuits; and I hope the period is not far distant when as much will be said for anatomists." Can I be mistaken, after the perusal of the last two passages, in agreeing with M. Cuvier on the advantage of finding such a collection of fossil remains within our reach, and from this accordance to deduce the equal advantage of having had such an intellect to explain, to apply, and to appreciate the evidences thus presented to man of the changes which have taken place in the earth which he inhabits?

I now have to notice the two editions of the Régne Animal, which, with the Tableau Elémentaire, I have already esteemed as one and the same work; the first edition being a completion of the sketch contained in the Tableau, and the second edition being an enlargement of the first, with a slight alteration in the classification, necessitated by the progress of discovery. Having used the dissecting knife through every

class of nature *, M. Cuvier was necessarily
struck with the confusion of systems, their want
of conformity to the internal structure of animals,
and the heap of synonymes which multiplied
species to infinity; and, as may be seen throughout this work, accustomed from the earliest age
to entertain elevated views, and to practise method, it was absolutely necessary, even for his
own future convenience, that he should rid classification of the incumbrances which impeded its
advancement. The manner in which he accomplished this object, is displayed in the preface to
the first edition of the Régne Animal, in the
most interesting manner, together with the assistance he received from his colleagues, especially his brother, M. Frederic Cuvier, whose
observations on the teeth of mammalia were of
the greatest service to him in forming some of
his minor divisions. This preface well describes
the state in which he found the classification of
animals when he first undertook to free it from
its shackles, and is annexed to both editions.
The great outlines of his system may be given
nearly in M. Cuvier's own words : — " There

* One of M. Cuvier's most able assistants in the dissecting
department was M. Rousseau.

exist in nature four principal forms, or general
plans, according to which all animals seem to
have been modelled, and the ulterior divisions of
which, whatever name the naturalist may apply
to them, are but comparatively slight modifications, founded on the developement, or addition
of certain parts, which do not change the essence of the plan." The introduction to these
volumes contains the definition of classes, orders,
genera, &c., a general view of that which is
called organisation, particularly that of animals,
its chemical composition, its forces, its intellectual and physical functions, and the application of method to the four great forms of the
animal kingdom.   From the latter I must be
allowed to make a short extract.   " In the first
(form), which is that of man, and the animals
which most resemble him, the brain and the
principal trunk of the nervous system are enclosed in a bony envelope, which is composed of
the skull and vertebræ : to the sides of this middle column are attached the ribs and bones of
the limbs ; all of which form the frame-work of
the body.   The muscles which give action to
these bones generally cover them, and the viscera are contained in the head, and the trunk,
or body.   These are styled vertebrated animals :

they all have red blood, a muscular heart, a
mouth with two jaws, one above, or before the
other, distinct organs for sight, hearing, smell,
and taste, placed in the cavities of the face,
never more than four limbs, the sexes always
separated, and a similar distribution of medullary masses, and of the principal branches of the
nervous system. When thoroughly examining
each of the parts of this great series of animals,
we shall always find some analogy between them
all, even in the species the farthest from each
other; and we can follow the gradations of the
one same plan, from man to the last of the
fishes. In the second form there is no skeleton,
the muscles are only attached to the skin, which
forms a soft envelope, contractile in various
senses, in many species of which are engendered
stony plates, called shells, the position and production of which are analogous to those of the
mucous body to which they belong. Their nervous system and viscera are contained in this
general envelope; the former is composed of
several scattered masses, united by nervous
threads, the principal of which, placed on the
œsophagus, bear the name of brain. In general,
they only possess the senses of taste and sight,
and even the last is often wanting. Only one

family can boast of the organ of hearing; they
have always a complete system of circulation,
and organs peculiarly adapted to respiration;
those of digestion and secretion are nearly as
complicated as the same organs in vertebrated
animals. This second form is called that of
molluscous animals; and although the general
plan of their organisation is not as uniform,
with regard to their external appearance, as that
of vertebrated animals, there is still a greater or
lesser degree of resemblance in the structure
and functions of these parts.

" The third form is that which is to be found
in insects, worms, &c. Their nervous system
consists of two long cords, which traverse the
belly lengthwise, and are enlarged from space to
space into knots, or ganglions. The first of
these knots is situated above the œsophagus, and
is considered as the brain; but it is scarcely
larger than those which are in the belly, with
which it communicates by threads, which embrace the œsophagus like a collar. The envelope
of this structure is divided by transversal folds
into a certain number of rings, the teguments of
which are sometimes hard, and at others soft,
but to the interior of which the muscles are
always attached. The trunk often bears ar-

ticulated members on its sides, but is as often without. These are the articulated animals, and it is among them that we observe the passage of the circulation in closed vessels, or nutrition by imbibition, and the corresponding passage of respiration in the circumscribed organs called tracheæ, or aërial vessels spread over the whole of the body, by means of which it is performed. Like the second form, there is but one family which possesses the organs of hearing, and those of the taste and sight are chiefly developed. If they have any jaws they are always lateral. The fourth form embraces all the animals known under the name of zoophytes, and is called that of radiated animals. In all the preceding, the organs of movement, and the senses, are symmetrically disposed on the two sides of an axis; they have a posterior, and an anterior face of dissimilar appearance. But in those now mentioned, they are as if composed of rays round a centre, even when there are but two series of these rays, for then the two faces are alike. They approach the homogeneity of plants: they have no very distinct nervous system, nor particular organs for the senses. In some there are scarcely any vestiges of circulation; their respiratory organs are almost always on the

surface of their bodies; the greater number have
but one bag without issue for an intestine, and
the last families only present a sort of homogeneous pulp, movable, and sensible to the
touch." Here I must again impress on the
reader, that M. Cuvier's first great discovery
was the necessity of separating this last form of
animals from the general mass of insects and
worms, having read his Memoir, pointing out
the characters and limits of mollusca, echinodermes, and zoophytes, to the Society of
Natural History in Paris, on the 10th of May,
1795. From this he ascended to animals of more
complicated form, for it is only a man of narrow
mind that can treat any part of natural history
with contempt. All others will see in it "a continuance of that command given to Adam, to
see, to name, and to use the creatures put under
his control." No branch of it, however trifling,
but may be ennobled by the manner in which it
is pursued; and when the student carries all its
wonders back to the one Great Source, the
smallest worm and the most beautiful of his
own species will afford him subjects for the
deepest contemplation.

The Régne Animal begins with that being
which most interests us, of which there is but

one genus, and one species; the differences we
observe in him being but varieties, which are
termed races. Nothing can be more calculated
to excite profound attention than M. Cuvier's
definition of Man, and it would be so much in-
jured by selecting passages from it, that extracts
can only be made from that portion entitled
" Varieties of the Human Race." — " Three of
these are eminently to be distinguished from
each other; the White or Caucasian, the Yellow
or Mongolian, the Negro or Ethiopian. The
Caucasian, to which we (Europeans) belong, is
remarkable for the beautiful oval form of the
head, and from it have proceeded those people
who have attained the greatest civilisation, and
have held dominion over the rest. It varies in
complexion, and the colour of the hair. The
Mongolian is recognised by its prominent cheek
bones, flat face, narrow oblique eyes, straight
black hair, scanty beard, and olive tint. From
it have arisen the great empires of China and
Japan, and by it some great conquests have been
achieved, but its civilisation has always remained stationary. The Negro race is confined
to the south of the Atlas chain; its complexion
is black, hair woolly, skull compressed, nose
flattened, muzzle projecting, lips thick, and

nearly approaches monkies. The natives which compose it have always remained in a comparatively barbarous state.

" The Caucasian race is subdivided into three great branches, and is supposed to have had its first origin in that group of mountains situated between the Caspian and Black Seas. The Syrian branch spread to the south, and produced Assyrians, Chaldeans, Arabs, Phenicians, Jews, Abyssinians, and probably Egyptians. From thi branch, always inclined to scepticism, have arisen the religious doctrines most generally adopted. Sciences and letters have sometimes flourished among them, but always under some strange shape, or in some figurative style. The Indian, German, or Pelasgic branch took a still wider range, and the affinities of its four principal languages are more multiplied.—The Sanscrit, which is still the sacred language of the Hindoos, is the parent of most of the Hindostanee tongues. The Pelasgic was the source whence came the Greek, Latin, and present dialects of the south of Europe. The Gothic or Teutonic, whence are derived the north and north-west languages, such as German, Dutch, English, Danish, Swedish, and their varieties; and, lastly, the Sclavonic, whence came the languages of the

I

north-east, viz. the Russian, the Polish, the Bohemian, and the Vendean. It is this great and respectable branch of the Caucasian race which has carried philosophy, science, and art to their greatest perfection, and of which it has been the depositary for thirty centuries. The inhabitants of the north, such as the Samoyedes, the Laplanders, and the Esquimaux, come, according to some, from the Mongolian race, and according to others they are the degenerated off-spring of the Scythian and Tartaric branch of the Caucasian. The Americans cannot be clearly brought back to either of our races of the Old World; and yet, nevertheless, they do not possess a sufficiently precise and constant cha-racter to form a peculiar race. Their copper complexion is far from being enough; their black hair and their beard would approach them to the Mongolian, if their marked features, their nose, equally projecting with our own, their large and open eyes, did not oppose this idea, and assimilate them to our European forms. Their languages are as innumerable as their nations; and no one has as yet been able to seize on demonstrative analogies between them-selves, or between them and the inhabitants of the ancient Continent."

The second order of Mammalia, is that of
the Quadrumana, or apes, who are many of
them men without reason : the third contains
the Carnivora, which affords lions, tigers, &c.
and all that we can imagine of fearfulness and
ferocity; and yet, from whence we derive our
faithful dogs, our domestic cats, and our most
beautiful furs.   The fourth is named Marsu-
pialia, and consists of those singular animals
whose young are prematurely born, and take
refuge afterwards in a pocket attached to the
body of the mother, till they are able to take
care of themselves. The fifth, Rodentia, is that
in which we find squirrels, rats, beavers, hares,
&c.   The sixth, Edentata, furnishes us with
that disgusting animal the sloth, and the ornithorynchus, that extraordinary native of New
Holland, which has a beak like that of a duck,
feet so webbed as to resemble fins, fur like that of
a weasel, and which has by some been supposed
to lay eggs. The seventh order is called Pachy-
dermata, and in it we find the largest animals
which walk on the surface of the globe, such
as the elephant, the hippopotamus, the rhinoceros, and also the horse, which has been in all
ages the most easily adapted to the use of
mankind. The eighth, Ruminantia, whence

come the cow, the camel, and the reindeer;
the two latter of which convey their masters
over the hottest or the coldest regions of the
earth; and lastly, the ninth, or Cetacea, which
presents us with the mighty monsters of the deep.
These nine orders are subdivided into families,
genera, subgenera, &c., and the most important
species are noticed with considerable detail.

From Mammalia, M. Cuvier proceeds to
Birds; and after their physiological description,
he also divides them into orders, pointing out
the reasons of such divisions, and carrying us
through every portion of the winged tribe. He
first embraces the birds of prey, such as the
vultures, who act, as it were, the part of sca-
vengers; the eagles who prey by day, and owls
who thieve by night; the second contains the
numerous genera of the Passeres, they are not
so violent as birds of prey, properly so called,
nor have they the decided habits of the Gallinaceæ, or aquatic birds, but devour insects,
fruits, and grains; those who pursue insects
will also feed on smaller birds, and have slender
beaks; and those who eat grains have thick
beaks. The first subdivisions of this order depend on the feet, and the others on the form of
the beaks. Among them we find our singing

birds, our birds of paradise, and our humming
birds.   The third order is that of the Climbers,
such as the parrot, &c.   The fourth embraces
the Gallinaceæ, whence we derive our farm-
yard fowls, and most of our game.   The fifth,
or Grallæ, gives us the ostrich, the cassowary,
the sacred ibis, &c. ; and the sixth, named the
Palmipedes, presents us with ducks, geese,
pelicans, &c. &c.

As this first volume is conducted, so does the
Règne Animal lead us through every part of the
animal world, describing all in forcible and clear
terms, neither saying too much nor too little,
commenting upon whatever is most remarkable,
viewing the affinities of these beings according
to their just value, and giving a model for methodical arrangement, inasmuch as it approaches
as nearly as possible to nature.   It must be observed, however, of the third volume, that as the
considerable increase of Entomology, in common
with every other branch of natural history, rendered it impossible for one man, in a reasonable
time, thus minutely to treat the whole series of
life, M. Cuvier called in the assistance of M.
Latreille for that part of the work which relates
to Insects and Crustacea ; but where the reader
will find those enlightened views, and that beau-

tiful method, which is every where practised by his great colleague. " The principles on which M. Cuvier's divisions rest, will necessarily preside over all the changes which still more extended observation will render indispensable; but the basis of zoological classification is for ever laid, and its solidity will prove, better than all the discourses of future naturalists, the elevated genius of the author." *

The galleries of stuffed animals at the Jardin des Plantes, containing thousands of species, are all arranged according to the system of the above series, the writer of which desired no better than to lay before the world the reasons on which he founded it, and to give at the same time an equal opportunity for correction and admiration. Among the specimens there placed, are those which he amassed for the labour I have next to describe, many of which he had dissected with the most minute attention, and which increased this part of the collection to the amount of nearly five thousand species.

The great work on Ichthyology contains an application of M. Cuvier's principles to one peculiar branch of natural history, and was not

* Laurillard.

only intended by him as an example of the extent of which such an undertaking is capable,
but served the double purpose of aiding his
further researches among fossil fishes. It was
announced by himself in the conclusion of that
on Fossil Remains, in the following terms:—
" I shall now consecrate the remainder of my
time and strength to the publication of those
researches already made in the Natural History
of Fishes, but, above all, to the termination of
my general Treatise on Comparative Anatomy."
Scarcely did he seem to breathe between the
finished and the commenced undertaking; in
fact, the materials for several were collecting at
the same time; that which he termed his " General Treatise on Comparative Anatomy" was
always in preparation; every week brought a
fresh accumulation of notes and drawings; many
of the latter, and all of the former, made by his
own hand. The plan of the Ichthyology was
laid before the public by M. Cuvier, in a Prospectus describing the state of this branch of
the science, his actual resources, and those he
hoped to enjoy. M. Valenciennes, now Professor of Mollusca to the Museum of the Jardin
des Plantes, was called in to aid him in the innumerable details attendant on such an enter-

prise, and is now charged with the continuation
of the task which his great master left unfi-
nished. Eight volumes were published at the
time of M. Cuvier's death, and, since then, M.
Valenciennes has added another; the whole to
be completed in twenty volumes.*

The title at once implies the nature of what is
to follow: — " Natural History of Fishes, containing more than Five Thousand Species of these
Animals, described after Nature, and distributed
according to their Affinities, with Observations
on their Anatomy, and critical Researches on
their Nomenclature, antient as well as modern."
Linnæus determined 477 species, and De Lacépéde 1500; thus, without calculating on the
multiplication caused by the synonymes of these
authors, the increase made by M. Cuvier is
enormous. Throughout the work one species is
chosen from each group for detail, and that
preferred which is the most interesting, or the
easiest to procure. This is described with the
greatest minuteness, and serves not only as a

* This ninth volume was half printed during the life of
M. Cuvier; and he left, in manuscript of his own writing,
enough for three or four more volumes; but this being in
detached pieces, it will be scattered through the rest of the
work, according to the progress of the subject.

type, but a means of comparison for the charac-
teristic but simple differences between the other
species which compose the group. The necessity of stating the different names given by
various authors, and the discrimination required
to separate truth from fable in that which he
reported of their economy, demanded the exquisite judgment and profound experience which
rendered M. Cuvier so capable of the task; and
there was a general eagerness felt, which does
credit to naturalists and collectors of all countries, to offer to him every specimen, every discovery, every observation, even before the person
so offering had himself published the particulars.
This was the latest work of magnitude undertaken by M. Cuvier; and it is easy to judge, by
solely viewing the rapid growth of this one
branch, how every thing advanced under his
influence and his personal exertions, and how
materials poured upon him from those who were
sure of receiving justice from his hands, and
many of whom, rendered incapable by other
pursuits or circumstances of publishing their observations on their own account, were delighted
to be mentioned in his pages as among the very
humble contributors to his glory.

But in this publication, which is accompanied

by numerous and beautiful engravings, especially those made from the drawings of M. Laurillard, on the anatomy of the perch, we find a
new feature.   M. Cuvier becomes the historian
of that part of the science of which he treats;
and nothing can be more clearly or impartially
given than the progress of Ichthyology, from the
first certain glimpses to be met with concerning
its existence; and the place, the means, the
results, the influence of every labourer in the
cause, are set before us with wonderful precision
and order.   But as this is, with the exception of
the Memoirs on Mollusca (published at various
times in the Annales du Muséum, and now collected into one quarto volume), the only work
of M. Cuvier devoted to one single branch of
natural history, it may be interesting to give an
idea how it is conducted.   The history above-
mentioned forms, as it were, an introduction to
the whole, and concludes in these words: —
" As for us, the only wish we can now form, is,
that the work which we have undertaken may
not be found unworthy, either of the illustrious
writers whose labours we seek to continue, or
of the aid and encouragement we have received
from so great a number of friends, and from the
patrons of natural history.   Happy if we could

hope, in our turn, that our endeavours may rank among those which have marked the epochs of science. It is to this that all our efforts tend."

From the history, M. Cuvier proceeds to give a general idea of the nature and organisation of Fishes. The following is an extract from this part : — " Being aquatic, that is to say, living in a liquid which is heavier, and offers more resistance than air, their forces for motion have been necessarily disposed and calculated for progression, and elevation, which is also accomplished by them with ease. Hence arises that form of body which offers least resistance, the chief seat of muscular force residing in the tail, the brevity of their members, the expansibility of these members, and the membranes which support them, the smooth or scaly teguments, and the total absence of hairs or feathers. Breathing only through the medium of water, that is, for the purpose of giving an arterial nature to their blood, profiting by the small quantity of oxygen contained in the air, which is mingled with the water, their blood is necessarily cold, and their vitality, the energy of their senses and movements, are consequently less than in Mammalia and Birds. Their brain, therefore, or rather a composition similar to it, is proportionably

much smaller, and the external organs of their
senses are not of a nature to admit of powerful
impressions. Fishes, in fact, are, of all vertebrated animals, those which have the least apparent signs of sensibility. Having no elastic air
at their disposal, they have remained mute, or
nearly so, and all those sentiments awakened
or sustained by the voice have remained unknown to them. Their eyes almost immoveable,
their bony and rigid countenance, their members
deprived of inflexion, and every part moving at
the same time, do not leave them any power of
varying their physiognomy or expressing their
emotions. Their ear, enclosed on every side by
the bones of the skull, without external conch
or internal labyrinth, and composed only of a
few bags and membranous canals, scarcely
allows them to distinguish the most striking
sounds ; and, in fact, an exquisite sense of hearing would be of very little use to those destined
to live in the empire of silence, and around
whom all are mute. Their sight, in the depths
of their abode, would be little exercised, if
the greater number of the species had not, by
the size of their eyes, been enabled to supply
the deficiency of light; but even in these species, the eye scarcely changes its direction ; still

less can it change its dimensions, and accom-
modate itself to the distance of objects; its iris
neither dilates nor contracts, and its pupil remains the same in every degree of light. No
tear bathes this eye, no eyelid soothes or protects
it; and, in fishes, it is but a feeble representation
of that beautiful, brilliant, and animated organ
of the higher classes of animals. Procuring food
by swimming after a prey which also swims with
greater or lesser rapidity, having no means of
seizing this prey but by swallowing it, a delicate sense of taste would have been useless to
fishes had nature bestowed it on them. But
their tongue, almost immoveable, often bony, or
armed with dentated plates, and only receiving
a few slender nerves, shows us that this organ is
as little sensible as it is little necessary. Smell
even cannot be as continually exercised by fishes
as by animals which breathe air in a direct manner, and whose nostrils are unceasingly traversed
by odoriferous vapours. Lastly, we come to the
touch, which, from the surface of their bodies
being encircled by scales, by the inflexibility of
the rays of their members, and by the dryness of
the membranes which envelope them, has been
obliged to seek refuge at the end of their lips;

and even these, in some species, are reduced to a dry and insensible hardness."

In the whole of the chapter from which the above passage is selected, there is a poetical feeling, in which M. Cuvier rarely indulged when treating of science, but with which we find he could occasionally sport without injury to his subject. In the next chapter he resumes his more precise manner; and the contrast is the more striking, as this chapter may be almost styled a collection of aphorisms. It speaks of the exterior of fishes, and is succeeded by others containing the osteology, myology, brain and nerves, nutrition, reproduction, and a general summing up and methodical distribution of this class into its great divisions, its natural families, &c. From the latter may be selected a passage well calculated to prevent those who study systems from falling into a very common error. " Let it not be imagined, because we place one genus or one family before another, that we consider them as more perfect, or superior to another in the series of beings. He only could pretend to do this, who would pursue the chimerical project of ranging beings in one single line,—a project which we have long renounced. The more progress we have made in the study

of nature, the more we are convinced that this
is one of the falsest ideas that has ever resulted
from the pursuit of natural history; the more we
have been convinced of the necessity of considering each being, each group of beings, by
itself, and the part it plays by its properties and
organisation, and not to make abstraction of any
of its affinities, or any of the links which attach
it, either to the beings nearest to it, or the most
distant from it. Once placed in this point of
view, difficulties vanish, all arranges itself for
the naturalist: but systematic methods only
embrace the nearest affinities; and by placing a
being only between two others, they will always
be wrong. The true method is, to view each
being in the midst of all others: it shows all
the radiations by which it is more or less closely
linked with that immense network which constitutes organised nature; and it is this only
which can give us that great idea of nature, which
is true, and worthy of her and her Author; but
ten or twenty rays often would not suffice to
express these innumerable affinities .... We shall
therefore approach to each other those whom
nature has approximated, without feeling obliged to put into our groups the beings she has
not placed there; and making no scruple, after

having demonstrated, for example, all the species
which will admit of being arranged in a well-
defined genus, all the genera which may be
placed in a well-defined family, to leave out one
or several isolated species or genera, which are
not attached to others in a natural manner;
preferring the honest avowal of these irregu-
larities, if we may be allowed to call them so, to
those errors which must arise from leaving these
species, and anomalous genera, in a series, the
characters of which they do not embrace."

The first great division of Fishes treated of by
M. Cuvier, and with which the second volume
commences, is that of the Acanthopterygii, or
fishes with spinous rays to their fins; and foremost amongst these, is the numerous family of
the Perches, or Percoïdes, which occupies the
two succeeding volumes. The fourth volume
contains the family of the Joues Cuirassées, many
of which, and especially those of the tropical
seas, present themselves under extraordinary
and exaggerated forms, and to which belong
the beautiful little sticklebacks of our running
streams. The fifth volume embraces the Scien-
oïdes; the sixth, the Sparoïdes, and the Menides;
the seventh, the Squammipennes, and the Pha-
ryngiens Labyrinthiformes; and the eighth and

ninth, the Scomberoïdes. Each volume is closed
by the additions and corrections which the authors have found it requisite to make during the
progress of their publication ; and I have offered
this short list, because it has been a question
often repeated, even to myself, how far this noble
work was advanced when its progress was so
grievously arrested. It is the intention of M. Valenciennes to proceed as rapidly as possible with
the rest, designating those parts which are solely
due to the exalted genius, under whose auspices
he has become worthy of continuing thi. extensive and admirable enterprise.*

* I have always been very much struck with one part of
this work, and therefore cannot forbear calling the attention
of the reader to it.   It is the way in which M. Cuvier refutes
the opinions of M. Geoffroy St. Hilaire, who had long
opposed him with considerable warmth.   As far as relates to
Fishes, M. Cuvier, in notes at the bottom of certain pages,
places his antagonist's arguments in two columns, and by the
side of them, in two others, sets forth his refutations.   Not a
word of personal feeling is added, not a single argument is
brought in, to aid in persuading the reader that he is right ;
there are the two systems, equally exposed, and he who pe-
ruses them, perfectly at liberty to verify and judge for himself.
This difference of opinion being pursued with acrimony on several occasions by M. Geoffroy, it at last became a matter of
discussion before the Institute ; and M. Cuvier, who had long
remained silent with the most heroic forbearance, at length
was induced to reply.   After some little time, M. Geoffroy
retired from this direct contest ; but it is to be hoped, that

K

In noticing the Ichthyology, I have had occasion to speak of M. Cuvier as the historian of
the science to which he was devoted ; and this
leads me to mention here, the annual reports
made by him at the Institute, in which, from
the age of twenty-six, he had been accustomed
to lay before that body the labours of its members and correspondents,   thereby forming a
general history of science from that period till
his death.   In these " Analyses des Parties Physiques des Travaux de l'Académie des Sciences,"
we see the universality of his genius and acquirements ; and, like almost all his other undertakings,
we may consider this mass of reports, and the
qualifications necessary for the making of them,
as alone sufficient for the employment of a life.
They comprehend, first, Meteoro ogy and Natural
Philosophy in general ; secondly, Chemistry and
Physics, properly so called, and when the explanation of the facts did not demand calculation ;
thirdly, Mineralogy and Geology ; fourthly, Vegetable Physiology and Botany ; fifthly, Anatomy

---

the surviving friends of M. Cuvier will one day publish his
opinions separated from his great works, so that they may be
accessible to those who may not have either time or opportunity to seek them in the general tenor of his publications.

and Physiology; sixthly, Zoology; seventhly, Travels which were connected with the advancement of natural sciences: eighthly, Medicine and Surgery; ninthly, the Veterinary Art; and tenthly, Agriculture. From these analyses a just idea may be formed of most of the principal discoveries made in all these branches of science during the time of M. Cuvier; for not only did the members and appointed correspondents of the Institute feel it a duty to communicate their endeavours to this body, but many strangers felt a laudable pride in submitting their efforts to those who would be likely to appreciate them. All is described by M. Cuvier in his usually clear and forcible language, " frequently surprising even the author himself by the lucidity with which his own ideas and experiments are set forth, and sometimes creating in him new or different views of the subject which had long occupied his thoughts." * The same fearlessness of rendering justice marked these reports, as well as the other productions of the writer; and from their impartiality, their truth, and beautiful unity, they might have been supposed rather to have related to times long past, than to have been

* Dr. Duvernoy.

a record of the labours of contemporaries. Not
a word of his own opinions or feelings escapes
him; he mentions his own works with the most
perfect modesty and simplicity, and scrupulously
states, with invariable fidelity, every argument
brought forward, even against his own views
and sentiments.

Besides these annual reports, M. Cuvier was
charged by the Emperor with a new task, which
he thus announces in a letter to his friend M. Duvernoy: — " All my labours are nearly arrested
by a work demanded by the Emperor, the greater
part of which has devolved upon me as secretary
to the class (of natural sciences). It is a history
of the march and progress of the human mind
since 1789. You may suppose to what a degree
this is a complicated undertaking, respecting
natural sciences; thus I have already written a
volume, without having nearly reached the end;
but this history is so rich, there is such a beautiful mass of discoveries, that I have become
interested in it, and work at it with pleasure. I
hope it will be a striking dissertation on literary
and philosophical history; but above all things, I
endeavour to point out the real views which
ought to guide ulterior researches." It may be
considered as a work of the same nature as those

which I have just been describing, only infinitely
greater in extent, inasmuch as it embraces a
larger portion of time, and extends to those who
were not in the habit of communicating with
the Institute.

Napoleon had conceived the bold thought of
embracing, at one view, all that the general impulse towards learning and science had produced since the above period; and it may be
unhesitatingly affirmed, that the execution of
his wishes accorded with the elevated feelings
from which they sprang. It commences with
one of those introductions which always rank
among the highest efforts of M. Cuvier's genius;
in which he sets before us,—if I may be allowed
so to express myself,—the sublimity of science;
and is throughout remarkable for the extensive
views it takes, and its unflinching impartiality.
The following beautiful passage is among the
concluding pages, which pages contain a solicit-
ation for amendments and continued protection
on the part of the Emperor: — " To lead the
mind of man to its noble destination,— a knowledge of the truth, — to spread sound and wholesome ideas among the lowest classes of the
people, to draw human beings from the empire
of prejudices and passions, to make reason the

arbitrator and supreme guide of public opinion ; these are the essential objects of science. This is how she contributes to the advancement of civilization ; this is why she merits the protection of those governments, who, desirous of erecting their power on the surest foundation, form their basis on the common good." This report, and the " Analyses des Travaux," have been collected together as far as 1827, and published as a sup- plement to the " Œuvres complètes de Buffon," edited by M. Richard, and form two octavo volumes.

The active part taken by M. Cuvier, in con- junction with other savants, in the "Dictionnaire des Sciences Naturelles," and the influence of his name, were doubtless of infinite service to this valuable enterprise. His Prospectuses were quite as remarkable as any of his other produc- tions, and many writers applied to him for assist- ance in this respect. It was not, however, only when sought that he contributed his aid ; but, saying to a young author, " Let me see your Prospectus," and having seen it, adding, " let me arrange this for you," the next day, a page or two of eloquence would be ready for the press, which could not fail to produce a favourable impression of the forthcoming publication. That which announced the Dictionary I have just mentioned, rapidly exposes the history of science up to that time, and vouches for the pains taken by the contributors to its pages, that the extent to which science has lately carried her researches should be in every way gratified. Those great names with which M. Cuvier's has been so often associated in France and in England, are mentioned in the first pages, in a manner so in- teresting, and so satisfactory, that I cannot resist the pleasure of quoting his words. The extract is preceded by a view of the advantages which science received from the precepts of Bacon, and is as follows: — " Nevertheless, it is pro- bable that Natural History would not have so soon arrived at the brilliant condition for which it had been prepared by these wise precepts, had not two of the greatest men who adorned the last century concurred, notwithstanding the op- posite natures of their views and characters (or, perhaps, by this very opposition concurred), in causing its sudden and extensive growth. Lin- næus and Buffon, in fact, seem to have possessed, each in his own way, those qualities which it was impossible for the same man to combine, and all of which were necessary to give a rapid impulse to the study of nature. Both passionately fond

of this science, both thirsting for fame, both
indefatigable in their studies, both gifted with
sensibility, lively imaginations, and elevated
minds; they each started in their career, armed
with those resources which result from profound
erudition. But each of them traced a different
path for himself, according to the peculiar bent
of his genius. Linnæus seized on the distin-
guishing characters of beings, with the most
remarkable tact; Buffon, at one glance, embraced the most distant affinities: Linnæus,
exact and precise, created a language on purpose to express his ideas clearly, and at the same
time concisely; Buffon, abundant and fertile in
expression, used his words to develope the extent of his conceptions. No one ever exceeded
Linnæus in impressing every one with the beauties of detail with which the Creator has profusely
enriched every thing to which he has given life.
No one better than Buffon ever painted the
majesty of creation, and the imposing grandeur
of the laws to which she is subjected. The
former, frightened at the chaos or careless state
in which his predecessors had left the history of
nature, contrived, by simple methods, and short
and clear definitions, to establish order in this
immense labyrinth, and render a knowledge of

individual beings easy of attainment: the latter,
disgusted at the dryness of antecedent writers,
who, for the most part, were contented with giving
exact descriptions, knew how to interest us for
these objects by the magic of his harmonious
and poetical language.   Sometimes the student,
fatigued by the perusal of Linnæus, reposed
himself with Buffon ; but always, when de-
liciously excited by his enchanting descriptions,
he returned to Linnæus in order to class this
beautiful imagery, fearing, that without such aid
he might only preserve a confused recollection of
its subject; and doubtless, it is not the least of
the merits of these two authors, thus incessantly
to inspire a wish to return to each other, although
this alternation seems to prove, and in fact does
prove, that, in each, something is wanting. As unfortunately is but too often the case, the imitators
of Linnæus and Buffon have precisely adopted
the defects of each of their masters ; and that
which was in them but a slight shade in a magnificent picture, is become the principal character
in the productions of many of their respective
disciples.   Some have only copied the dry and
neological phrases of Linnæus, without recollecting that he himself only looked upon his system
as the scaffolding of an edifice of much greater

importance, and that in the special histories which
his numerous labours have permitted him to write,
he has not neglected a single thing which belongs
to the existence of the being which he describes.
Others have only admired the general views and
lofty style of Buffon, without remarking that he
only decorated a series of facts, collected with
the most judicious criticism, with these brilliant
ornaments ; and even that nomenclature, which
they affect to despise, is always established by him
with great erudition, after the most careful and
ingenious discussion." I close this extract with
a remark made upon M. Cuvier by M. Duvernoy,
who has also cited the above passage in an éloge
on his illustrious master, addressed to his disciples at Strasburgh. — " May we not say, after
this, that he who so well appreciated these great
men, who so happily found in the one, that
which was wanting in the other, knew how to
unite the excellencies of both in his own writings;
or rather, that his genius, in its originality, had
nothing incomplete, nothing which could make
us feel the want of the true method on one side,
nor the absence of general views on the other."

A list of the articles contributed by M. Cuvier to the above mentioned Dictionnaire will
be found among the catalogue of his works at

the end of this volume; but that headed "Nature"
is too important to be passed over in silence
here ; to remain unnoticed in memoirs especially
intended to set forth his opinions; for it contains
the clearest and most satisfactory refutation of
the reigning controversies that has ever been
published in a separate form; though what these
opinions were, may be gathered from every thing
he has written.

" The word Nature, like all abstract terms
which find their way into common language,
has assumed numerous and divers significations.
Primitively, and according to its etymology, it
means that which a being derives from its birth,
in opposition to that which it may derive from
art. . . . It is in the nature of an oak to grow for
three centuries, to have hard wood, to attain a
great size, &c. It is in that of a bird to raise
itself in the air, to distinguish distant objects,
&c. Man is by nature capable of education;
his nature is weak, inconstant, &c. Each individual may possess, physically or morally, its
own peculiar nature ; it may be feeble or vigorous, mild or passionate, &c.

" This word Nature is also extended to things
which are not born, to unorganised beings in
general, in order to designate the peculiar and

intrinsic qualities which they always possess. The nature of gold is to be heavy, yellow, and not liable to decomposition by air or humidity, &c. Thus taken in its most generic acceptation, the nature of a thing is that which makes it what it is—that which distinguishes it, which constitutes it—in a word, its essence: and it is thus that we speak even of the Being of beings, —of Him in whom, and by whom, are all things; and therefore the expression applied to God, and to his attributes, is a most improper term when applied to the vilest and most perishable bodies. But that which exists in the nature of each individual, exists also in each species, and each genus; and thus, rising from abstraction to abstraction, we at length arrive at the idea of a general nature of all things; this embraces the qualities common to all beings, and the laws of their mutual affinities; it is the nature of things, taken in its most abstract sense. Lastly, by a figure of speech, common to all languages, this term has been employed for the things themselves, for the substances to which these qualities belong. Nature then is, all beings, or the universe, or the world; and when considered as contingent and in opposition to the necessary Being, to God, it is called Creation. Nature, the world, creation, the whole of created beings, are, then, so many synonymes.

" But by another of those figures of speech to which all languages are prone, Nature has been personified; existing beings have been called the works of Nature, and the general affinities of these beings among themselves have been called the laws of Nature. The definitive result of these affinities, which is a certain constancy of motion, a certain fixedness in the proportion of the species; in short, the preservation, to a certain degree, of the order once established; has been entitled the wisdom of Nature. Lastly, the enjoyments afforded to sensible beings have taken the name of the bounty of Nature. Here, under the name of Nature, the Creator himself is evidently represented; they are his works, his cares, his wisdom, and his goodness, which are thus meant. Nevertheless, it is by thus con-sidering Nature as a being gifted with intel-ligence and will, but secondary and limited with regard to power, that we are able to say of her, that she unceasingly watches over the preserv-ation of her works, that she makes nothing in vain, that she effects all by the most simple methods, that she contributes to the cure of dis-eases, but that she is sometimes overcome by the

force of malady; and other adages; many of
which are only true in a very limited sense, and
in a very different manner from that which they
seem to offer at the first glance. . . . In proportion as the knowledge of astronomy, physics,
and chemistry has been extended, these sciences
have renounced the false reasoning which resulted from the application of this figurative language to real phenomena. Some physiologists
only have continued to use it, because the obscurity in which physiology is still enveloped,
renders it necessary to attribute some reality to
the phantoms of abstraction, in order to practise
illusion on themselves and others, concerning
their profound ignorance as touching vital motion.

" Nevertheless, this antient idea of an active
but subordinate principle, distinct from ordinary
forms, and the laws of motion which should
preside over organisation, and which should keep
it in order, still prevails, not only in language,
but in the systems of a great many writers, who,
although they allow the justice of the distinctions we have now made, yet suffer themselves
to be drawn unconsciously towards doctrines
which have no other foundation. Such are the
doctrines of the ' Scale of Nature,' the ' Unity

of composition,' and others similar to these, which have all been imagined in consequence of the belief in a Nature distinct from the Creator, and less powerful than he is, and which have no evident support, but in those fancied limits which they place to his power.

" That each effect may proceed from a cause, which cause is to be traced to an anterior cause; that in this manner all events, all successive phenomena, may be linked together; that there may be no interruption in the march of nature, and that we may, in this sense, compare her to a chain, all the rings of which are attached to and follow each other; is evident on the least reflection. That the beings which exist in the world are so constructed as to maintain a permanent order; that they have, consequently, sufficient for all their wants; that their action and reaction may exist in every place, and at every moment, as necessary for this permanency; that it may be the same with the parts of each being; the very maintenance of this order teaches us. Lastly, that in this innumerable multitude of different beings, each, taken apart, may find some which resemble it more than others, by their internal and external forms; that it may be the same with these, relative to a third set; and that, consequently, we may be able to

group near each being, a certain number of
other beings which approach it in different degrees; must necessarily be the case. But, that
we ought to apply to the resemblances of these
simultaneous beings, that which is true concerning the relation of successive phenomena and
events; that the forms of these beings necessarily constitute a series or a chain, so that the eye
may gradually pass from one to the other, without finding any gap, any hiatus; in short, the
existence of a continued and regular scale in the
forms of beings, from the stone to the man;
this is what our three concessions by no means
prove; this is what is not true, whatever eloquence may have been used in tracing the ima-
ginary picture. The philosophers who have
supported this system of a scale of beings, at
each interruption which is pointed out to them,
pretend, that if a step is wanting, it is hidden
in some corner of the globe, where a fortunate
traveller may one day discover it. Nevertheless, all regions, all seas, have been explored;
the number of species collected increases every
day; there are, perhaps, an hundred-fold more
than when these paradoxical opinions began to
be established, and none of the spaces are filled
up; all the interruptions remain; there is no-

thing intermediate between birds and other classes; there is nothing between vertebrated animals and those which have no vertebræ. The distinctions of true naturalists remain in all their force; the laws of the co-existence of organs, those of their reciprocal exclusion, remain unshaken. Each organised being has in concordance all that is necessary for its subsistence; each great change, in one organ, produces a change in others. A bird is a bird in all and every part; it is the same with a fish or an insect. We cannot even conceive a being which, having certain wants, has not the power of satisfying them; a being which could have a part of its organisation allied to another part, suited to a different being, an intermediate being, in fact, that which is called a passage.

" Each being is made for itself, and in itself is complete : it may resemble other beings, each equally composed of what is fit for it, but none can be composed with a view to another, nor to join it to a third by affinity of form ; and that which is true of the least plant, of the least animal; that which is true of the most perfect of animals, man ; of the little world, as the ancient philosophers called it, is necessarily not the less

L

true of the great world, the globe, and all its inhabitants. The beings which compose it, and which people it, contribute to its existence; they are necessary to each other, and to the whole; they have been so since this existence has subsisted; they will be as long as it shall subsist. The world is like an individual, all its parts act on each other: we can imagine other worlds more or less rich, more or less peopled, the preservation of which rests on other means; but we cannot conceive the present world deprived of one or several of the classes of beings which inhabit it, any more than the body of man deprived of one or several of its systems of organs.

" There is, then, in the world, as in the body of man, that which is necessary, and nothing more. What law could have obliged the Creator unnecessarily to produce useless forms, merely to fill up the vacancies in a scale, which is only a speculation of the mind, and which has no other foundation than the beauty which some philosophers discover in it? But in every thing beauty consists in relative fitness: the beauty of the world is formed by the happy concourse of beings which compose it, in their mutual preservation, and in that of the whole, and not in

the facility which a naturalist may find in arranging them into a simple series.

" Nevertheless, to the hypothesis of a continued scale in the forms of beings, other philosophers have added that in which all beings are
modifications of one only; or, that they have
been produced successively, and by the developement of one first germ; and it is on this
that an identity of composition for all has been
engrafted. . . This system (as it now exists) seizes
hold of some partial resemblances, without having any regard to differences ; it sees in the worm
the embryo of the vertebrated animal; in the
vertebrated animal with cold blood, the embryo
of the animal with warm blood; it thus makes
one class spring from the other; they are but
different ages of one only; and the whole of
animal life has the same phases as the most
perfect individual of its species. From this naturally arises the consequence that, taking the
superior classes in an embryo state, we ought
there to find the inferior parts, and that the
composition of all must be alike, except the
greater or lesser developement of certain parts.
But these affinities, which offer something like
plausibility when announced in general terms,
vanish directly they are detailed, and a com-

parison is made, point by point. There is not less hiatus in the affinities of parts than in the scale of beings; in vain, in order to escape conviction, arbitrary suppositions are brought forward in the overthrow of organs incompatible with the links which attach them to the rest of the body; in vain, as a last resource, is figurative language (which no logic can penetrate) made use of; they are obliged to confess that certain parts, often numerous, are wanting in certain beings, without any apparent reason for their absence, other than because they did not agree with the whole of the being; and if in these pretended theories we seek a rational and general basis, what is to be found except the supposition of a nature limited in her mode of action?

" In fact, if we look back to the Author of all things, what other law could actuate him than the necessity of according to each being, whose existence is to be continued, the means of insuring that existence; and why could he not vary his materials and his instruments? Fixed laws of co-existence in organs were then necessary, but that was all; for, to establish others, there must have been a want of freedom in the action of the organising principle, which

we have shown to be mere chimera. In vain do
they have recourse to that other axiom, of being
obliged to make every thing by the most simple
means. Very far from its being more simple to
employ the same materials for different objects,
it is easy to conceive some instances in which
this method would have been the most complicated of all; and certainly nothing is less satis-
factorily proved than this constant simplicity of
means. Beauty, richness, abundance, have been
the ways of the Creator, no less than simplicity.

" Whenever they who, in recent times, have
sought to give a new form to the metaphysical
system of pantheism, and which they have entitled ' Philosophy of Nature,' have adopted
the two hypotheses of which we have just
spoken, they have added a third, quite of the
same kind. Not only each being, according to
these, represents all others, but it has a representation of itself in each of its parts. The
head is a complete body; the skull, composed
of vertebræ, is the spine; the nose is the thorax;
the mouth the abdomen; the upper jaw the arms,
the lower the legs; the teeth are fingers or
nails; and in this thorax, in these four members, are to be found the larynx, the ribs, the

shoulder-blades, and the basin, in a word, all the
bones.

"We comprehend, in fact, that those who
admit but of one single substance, of which
every individual existence is but a manifestation,
would have pleasure in adopting the idea that
these manifestations succeed each other in a
regular and progressive order; that they all
bear the impression, and, in some measure, become the images of one common type, or essential substance, and that each part, each part of
a part, not only represents the special whole
which contains it, but even the great whole
which contains all others. . . . .

"We, however, conceive nature to be simply
a production of the Almighty, regulated by a
wisdom, the laws of which can only be discovered by observation; but we think that these
laws can only relate to the preservation and harmony of the whole; that, consequently, all
must be constituted in a manner that contributes
to this preservation and to this harmony, but we
do not perceive any necessity for a scale of beings, nor for an unity of composition, and we
do not believe even in the possibility of a successive appearance of different forms; for it appears
to us that, from the beginning, diversity has

been necessary to that harmony, and that preservation, the only ends which our reason can
perceive in the arrangement of the world."

Besides the " Dictionnaire des Sciences Naturelles," there was yet another work of the same
kind to which M. Cuvier was a contributor— the
" Dictionnaire des Sciences Médicales." The
most important of the papers thus destined is
that headed " Animal ; " in which, after stating
that the power of will can only produce the
movements for which the body is adapted, and
that, consequently, the energy of the signs which
it gives will bear a proportion to the greater or
lesser perfection of the envelope, he takes a
rapid view of the beings which fill the interval
between the sponge, the *animality* of which
consists solely in the power of contraction ; and
the dog, or elephant, each of whom is gifted
with a sentiment which often bears the appearance of reason.

I shall confine myself to the extract of that
part which describes the lower order of animals,
having already, when mentioning other writings,
had occasion to speak frequently of the higher
classes. " A little above the sponges are the
monades, and other microscopic animals of an
homogeneous substance, simple and uncertain in

form, but which move in water with greater or
less rapidity. The polypes only exceed these
by having an invariable figure, and some distinct members round the mouth; several of
them, fixed to the solid masses which they themselves produce, have no motion but in their
members, and are incapable of changing place.
The radiata, or sea-nettles, ascend in organisation, by having several ramifications of the intestinal canal. The echinodermes possess an
envelope more or less hard, and their numerous
members serve them for progressive motion. At
this point the star form disappears, and gives
place to the symmetrical, where similar parts are
disposed along a line or axis. Almost all of the
most simple of these, the intestinal worms, live in
other animals; they have neither members, nor
heart, nor blood vessels; their body is elongated,
and sometimes articulated." To these succeed
insects, &c. &c. and the whole concludes with a
comparison between plants and animals.

It is not the just appreciation of Linnæus and
Buffon only that we owe to M. Cuvier: there
is yet another celebrated writer, whose real value·
may be gathered from his labours; and the profound learning evinced in the notes to M. Le-
maire's edition of Pliny show, that M. Cuvier

could make even his classical attainments serve
the science to which he was devoted. Pliny has
frequently been magnified into a great author
concerning natural history, and his writings
appealed to as a most indisputable source of
information. It seems, however, that he was
but a skilful compiler; he copied what others
had said before; he asserted many things from
common report, and could by no means be
relied on with that security which is due to
the naturalist who describes from personal observation. Thus, although there is much in him
to believe and to admire, considerable caution is
requisite in the study of his pages ; and it is
a most important service rendered to the inex-
perienced, to have identified the animals of
Pliny, to have shown how much is worthy of
confidence, and what should be rejected.

I am now about to notice a work of a very
different character from any which have hitherto
been presented: it is a very small duodecimo
volume of eighty-nine pages, but it is a gem
which owes nothing of its lustre to its size, and
sparkles, amid other brilliants, from the exquisite feeling which breathes in every line. It does
not delight us by the charms of its eloquence,
so much as by placing M. Cuvier before us as a

moralist, who derives his precepts from that
pure light which shines on all who seek it. The
subject is the distribution of the prizes founded
by M. de Montyon for virtuous actions. This
philanthropist had spent a life of usefulness, and
particularly sought to ameliorate the condition
of the lower classes, " that class of beings,
which," to use M. Cuvier's expressions, " he
saw exposed to poverty and disease; forced to
undergo severe and painful, even dangerous and
unhealthy labours; almost entirely deprived of
education; particularly open to the seductions
of vice, the torrent of passions, and brutal pleasures; often obliged to listen to the suggestions
of want and hunger, and having no resource
against these temptations in mental acquirement, in the habit of reflection, in public esteem,
in the hope of a better fate, or that ease of circumstances, which in other conditions is acquired by labour and good conduct."

M. de Montyon left legacies to hospitals; and
thinking, that after quitting these asylums in
too weak a state to work, the poor needed still
further aid, he destined a certain sum to this
purpose. Besides this, he left funds for bestowing prizes on those who invented machines to
be used in agriculture or mechanical arts, and

also on any one who should discover efficacious
remedies for the diseases which afflict humanity,
or diminish the danger to which workmen are
exposed in carrying on several of their occupations : he founded a third prize for books, which
should instruct the poor in moral conduct and
proper deportment; and, lastly, he instituted
that of virtue, exclusively in favour of the poorer
classes. This prize is annually bestowed, and
awarded by the Académie Française. In 1829
M. Cuvier was appointed, at the meeting of St.
Louis, to inform the public how the prizes had
been bestowed; and his discourse on the subject
forms the volume of which I now speak.

In his introduction to the history of those
who have obtained the prizes, the author says,
" Let us first ask the question, What is virtue?
An ancient philosopher answers, ' Remarkable
and brilliant virtue is that which supports woe
and labour, or which exposes itself to danger, in
order to be useful to others, and that without
expecting or desiring any recompense.' The
philosopher has well said that this is rare and
brilliant virtue; perhaps it is even above humanity; but let us observe, that its two principal characters are, usefulness to others, and
perfect disinterestedness. But we will turn from

pagan antiquity, open the Gospel, and there
seek an answer to the question proposed. We
read in the Holy Writings, ' Love God above
all things, and your neighbour as yourselves:
the law and the prophets are contained in these
two precepts.' Thus, he who has followed these
precepts is virtuous; he will have accomplished
the entire law. Now, what is it to love God?
How can we prove that we love him? It is by
conforming to his will, by doing that which he
orders; and the first thing which he commands
us to do, after loving him, is to love our neighbour as ourselves; and our neighbours are all
men, without distinction or exception, as the
Gospel also teaches us in the parable of the Sa-
maritan. This command, given us by God, has
been rendered easy and pleasing in execution by
himself having implanted in our souls, at our
birth, a love for our neighbours, a natural disposition to love our fellow-creatures, to rejoice in
their joy, and weep for their sorrows. This sym-
pathy, this soothing feeling, which religion calls
charity, is to be found in all pure and unper-
verted hearts, though it is not equally developed,
equally energetic in all. We feel that which we
owe to each other, not only justice, but succour
to the extent of our ability. Do not to others

that which you would not they should do unto
you ; and do to others as you would they should
do unto you. These are very simple rules, to
be comprehended even by children, and recognised by them as equitable and necessary ; they
are the foundation of all morality, and why are
they not always followed ? It is because we are
blinded by our passions, our inclinations, and
our interests. We have just said that God has
given us the feeling of love towards our neighbours, but he has also given us a love for ourselves, for our own preservation ; this sentiment
is not less natural than the other, and is not
wrong, because it is necessary ; it even teaches
us several virtues, such as temperance for the
sake of health, prudence to avoid danger, and
courage for the means of extricating ourselves
from difficulty. God tells us to love our neighbour as ourselves, that is, to tell us to love ourselves ; but when this love of self is carried to
excess, then it is that it merits the odious appellation of egotism ; then it prompts us to sacrifice others to ourselves, to wish to enrich
ourselves by their losses, to reckon others as nothing when our own satisfaction is concerned ;
then does it become a guilty feeling ; then does
it lead us to injustice and crime. It is even sad

and foolish to love ourselves only; and if we
have never done any thing for others, how
can we expect gratitude and help from them?
' C'est n'être bon à rien, de n'être bon qu'à soi.' *
He who stifles in himself the feeling of compas-
sion, and only obeys the dictates of self-love, is a
dangerous being in society, and who ought to be
reproved and punished in society, at least by
contempt. We may say, that almost all the evil
we commit arises from egotism; whilst the
greater part of our good actions is inspired by
love for our fellow-creatures. Therefore, the
best system of education is that which teaches
us to direct and control our self-love within its
just limits, and, at the same time, tends to develope and augment our love for others, our desire
of being useful, and doing them good. These
reflections lead us back to M. de Montyon, who,
always animated by this desire, wished to render
all men wiser, better, and happier. It was with
this intention that he founded the prizes of virtue,
the distribution of which has been confided to the
Académie Française, and this is the tenth time
of fulfilling this honourable mission. . . . But the
liberality of M. de Montyon, though great, must

* It is to be good for nothing to be only good to one's self.

be limited; and a choice must be made among those who are presented to us, each with the strongest claims. It may be imagined how difficult it is to make this choice: how painful, and even afflicting, it is to the judges to be obliged to compare, and coolly weigh actions which amount to sublimity; and, while animated to enthusiasm, or moved almost to weakness, thus impartially and calmly to pronounce judgment. Besides, what man can flatter himself that he can be exempt from error in such decisions? God alone is the judge of virtue, because he alone can read our hearts, penetrate into our motives, and know our intentions: God alone gives to virtue its real reward. We can only see the exterior, and presume on the motives, which we are bound to consider as pure and upright, when the actions bear the appearance of disinterestedness and goodness."

Thus far I have attempted, by translation, to give some idea of this beautiful little volume: but as the account of M. Cuvier's works draws near to the close, it will be desirable, occasionally, to give specimens of his style, by extracts from the French; and having thus stated the motive, these passages will be introduced whenever they seem to me to be best calculated for

displaying his powers. The following is selected
from the " Prix de Vertu * : "—" Nous avons à ra-
conter les bonnes actions de quinze autres per-
sonnes, auxquelles l'académie a décerné des
médailles ; au moment de commencer ces récits,
nous éprouvons une crainte, celle de fatiguer nos
lecteurs par la monotonie, et le défaut de variété,
ces récits vont se ressembler entre eux ; ce sera
toujours de la charité, toujours de la bienfaisance,
toujours un devoûment désintéressé aux infortunes
d'autrui ; et puis, il faudra toujours louer, toujours
admirer : ce n'est pas le moyen de réveiller et
de soutenir l'attention ; l'éloge nous fatigue ou
nous endort ; un écrivain Anglais dit spirituelle-
ment, que tous les panégyriques semblent confits

---

* We have now to recount the good deeds of fifteen other
persons, to whom the Academy has decreed medals ; but as
we commence these recitals, a fear assails us of fatiguing our
readers by monotony, by want of variety. These histories are
all alike ; it is always charity, always benevolence, always a
disinterested devotion to the cause of the unfortunate. These
we must for ever praise, for ever admire ; and this is not the
way to rouse or to fix attention. Eulogium fatigues or sends
us to sleep, and an English writer wittily says, that all pane-
gyrics seem to have been cooked in poppy juice. We however will abstain from saying a single word which may appear
to be given for the purpose of impressing these affecting circumstances. Still more forcibly do they carry their own recommendation with them ; and those who are so unhappy as
not to feel them, will not be capable of comprehending any
eulogiums which we could add to them.

dans du jus des pavots. Eh bien! nous nous abstiendrons de dire un seul mot qui pourrait sembler destiné à faire valoir des actions si touchantes; elles se recommandent assez par elles-mêmes; et ceux qui auraient le malheur de n'en être pas attendris, ne seraient pas même en état de comprendre les éloges que nous pourrions y ajouter."

One of the great prizes awarded on this occasion was five thousand francs to Louise Scheppler, whose history will, if I mistake not, be acceptable to the reader, as given by the Baron Cuvier. " Louise Scheppler has, perhaps, carried this industrious beneficence still farther, for it is not one family, it is an entire country which enjoys the fruits of her benevolence; a whole country which has been vivified by the charity of a poor servant. In the rudest part of the chain of the Vosges mountains is a valley, almost separated from the rest of the world. Sixty years back it afforded but scanty nourishment to a half-civilised population, consisting of only eighty families, distributed in five villages. Their ignorance and their poverty were equally great; they neither understood German nor French; a patois, unintelligible to any but themselves, was their sole language;

and, what is scarcely credible, their misery had
not softened their manners. These peasants,
like the lords of the middle ages, governed by
force, hereditary feuds divided families, and
more than once gave rise to acts of criminal
violence. A pious pastor, named John Frederick Oberlin, who has since become so celebrated, undertook to civilise them; and for this
purpose, like one who knew mankind, he first
attacked their poverty. With his own hands he
set the example for all useful labours, and, armed
with a pickaxe, he directed them in the construction of a good road, digging and labouring
with them; he taught them to cultivate the po-
tatoe; he made them acquainted with good
vegetables and fruits; showed them how to en-
graft, and gave them excellent breeds of cattle
and poultry. Their agriculture once perfected,
he introduced manufactures among them, in
order to employ superfluous hands; he gave
them a saving-bank, and put them in communi-
cation with the commercial houses of the neighbouring towns. As their confidence increased
with their improvement, he, by degrees, gave
them instruction of a higher nature. He himself
was their schoolmaster, till he could form one
capable of seconding his endeavours. Having
once learned to love reading, every thing became

easier; chosen works were brought to them to
aid the conversation and example of the pastor;
religious feelings, and, with them, mutual benevolence, insinuated themselves into their hearts;
quarrels, crimes, and lawsuits disappeared; and,
if by chance, some dispute arose, they, with one
accord, came to Oberlin, and begged him to put
an end to it. In short, when this venerable
man was nearly at the end of his career, he was
able to say, that in this province, once so poor
and thinly populated, he left three hundred families, regular in their habits, pious and enlightened in their sentiments, enjoying remarkable
ease of circumstances, and provided with the
means of perpetuating these blessings. A young
female peasant from one of these villages, named
Louise Scheppler, though scarcely fifteen years
of age, was so forcibly impressed with the virtues
of this man of God, that, although she enjoyed
a small patrimony, she begged to enter into his
service, and take a part in his charitable labours. From that time she never accepted any
wages; she never quitted him; she became his
help, his messenger, and the guardian angel of
the rudest huts. She afforded the inhabitants
every species of consolation; and in no instance
can we find a finer example of the power of feel-

ing to exalt the intelligence. This simple village girl entered into the elevated views of her
master, even astonishing him by her happy suggestions, which he unhesitatingly adopted in his
general plan of operation. She it was who remarked the difficulty that the labourers in the
fields experienced, in combining their agricultural employments with the care of their younger
children, and who thought of collecting together,
even infants of the earliest age in spacious halls,
where, during the absence of their parents, some
intelligent instructresses should take care of,
amuse, teach them their letters, and exercise
them in employments adapted to their ages.
From this institution of Louise Scheppler arose
the infant schools of England and France, where
the children of the working classes, who would
otherwise be exposed to accidents and vicious
examples, are watched over, instructed, and protected. The honour of an idea which has produced such beautiful results is solely due to this
poor peasant of Ban de la Roche; to this she
consecrated all her worldly means, and, what are
of more value, her youth and her health. Even
now, though advanced in years, she, without receiving the smallest compensation, assembles a
hundred children round her, from three to seven

years of age, and instructs them according to
their capacities.   The adults, thanks to M. Oberlin, have no further moral wants; but there are
yet some, who in sickness or old age have need
of physical aid.   Louise Scheppler watches over
them, carries them broth, medicine, in short,
every thing, not forgetting pecuniary succour.
She has founded and regulated a sort of Mont
de Piété *, of a peculiar kind, which would be an
admirable institution elsewhere, if it could be
multiplied like the infant schools; for it is
among the very small number of those which
merit the name given to them, for money is
there lent without interest and without securities.
When M. Oberlin died, he, by will, left Louise
Scheppler to his children; the simple words of
a dying master may be heard with interest, and
will be more eloquent than any thing we can
add: — ' I leave my faithful nurse to you, my
dear children, she who has reared you, the indefatigable Louise Scheppler; to you also she
has been a careful nurse, to you a faithful mo-

* The Mont de Piété of Paris, managed by a company of
individuals, was first established on the same principle as that
of Louise Scheppler, but is now the general establishment for
pawning, to which all the minor pawnbrokers of that city
belong.

M 3

ther and instructress ; in short, every thing : her
zeal has extended still further; for, like a true
apostle of the Lord, she has gone to the villages
where I have sent her, to gather the children
round her, to instruct them in the will of God,
to sing hymns, to show them the works of their
all-powerful and paternal Maker, to pray with
them, to communicate to them all the instruc-
tions she had received from me and your own
excellent mother. The innumerable difficulties
she met with in these holy occupations would
have discouraged a thousand others ; the surly
tempers of the children, their patois language,
bad roads, inclement weather, rocks, water,
heavy rain, freezing winds, hail, deep snow, nothing has daunted her. She has sacrificed her
time and her person to the service of God.
Judge, my dear children, of the debt you have
contracted to her for my sake. Once more, I
bequeath her to you ; let her see, by your cares,
the respect you feel towards the last will of a
father, — I am sure you will fulfil my wishes,
you will in your turn be to her all together,
and each individually, that which she has been
to you.' Messieurs and Mesdemoiselles Oberlin,
faithful to the wishes of their father, were desirous of bestowing on Louise the inheritance of

a daughter; but nothing could induce this generous woman to lessen the small patrimony left
by her master; and all she asked was, permission to add the name of Oberlin to her own.
Those who claim this honourable appellation as
a birthright, think themselves still further honoured by her sharing the title."

In his office of Secretary to the Academy of
Sciences, it was also the duty of M. Cuvier to
read an éloge upon the deceased members of
that body before a public meeting. As his
peculiar department did not extend to the calculating sciences, the labours of those who devoted themselves to such devolved upon the
other secretary; but all the éloges written by
M. Cuvier have been collected at various times,
and published in successive volumes. Before I
give an account of them, a few remarks upon
his delivery may be desirable. The very slight
accent of Montbéliard which might be traced in
his conversation, entirely disappeared while reading or speaking in public; his voice could be
heard every where without being pitched in too
elevated a key, his articulation was remarkably
clear and distinct without being affected, so that
foreigners found it easier to comprehend him
than most of the French orators, and there was

a tone of feeling, a certain play of countenance, which carried his auditors with him in all the sentiments he tried to inspire. There was nothing in the least declamatory or theatrical, in order to arrest the attention; but his melodious tones, his elegant turn of expression, and natural grace of manner, gave a charm to the shortest phrases. These last perfections were so much the more remarkable, as emphasis was the fashion in academical discourses when he commenced his career, and it was like creating a new school to return to nature.

I now resume the description of the éloges, which form three volumes in octavo; and, as several remain which have only been published for the members of the Institute, it is to be hoped that, ere long, a fourth volume will be added. The first contains, previous to the éloges, "Reflections on the Progress of Science, and its Influence on Society," read at the first annual sitting of the four academies. I must stop here to cite a most eloquent sketch from it, which leads us from the first helpless state of man to his present powerful condition, for it will give to my readers a proof of M. Cuvier's power of bringing important truths before us by one luminous flash from his pen.

*" Jeté faible et nu à la surface du globe, l'homme paraissait crée pour une déstruction inévitable : les maux l'assaillaient de toute parte, les remèdes lui restaient cachés ; mais il avoit

---

* Man, who had been thrown on the surface of the globe in a state of feebleness and nakedness, would appear to have been created for inevitable destruction : evils assailed him on all sides, and the remedies for them appeared to be hidden from him ; but he had been endowed with talents for their discovery. The first savages gathered nourishing fruits and wholesome roots in the forests, and thus conquered their most pressing wants. The first shepherds perceived that the stars followed a regular course, and by them directed their steps across the desert. Such was the origin of physical and mathematical sciences.

No sooner had the genius of man ascertained that it was possible to combat nature by her own means, than it no longer rested ; it watched her incessantly, and continually gained new conquests over her, each marked by some amelioration in the state of society. Then succeeded, without interruption, those meditating minds, which, being the faithful depositaries of acquired doctrines, were constantly occupied in connecting them, in vivifying the one by the help of the other, and which have led us, in less than forty centuries, from the first attempts of these pastoral observers, to the profound calculations of Newton and Laplace, to the learned enumerations of Linnæus and Jussieu. This precious inheritance, always augmenting, borne from Chaldea to Egypt, from Egypt to Greece, hidden during ages of misery and darkness, unequally spread among the people of Europe, has been every where followed by riches and power ; the nations who have welcomed it, have become mistresses of the world, and those who have neglected it have fallen into feebleness and obscurity.

reçu le génie pour les découvrir. Les premièrs
sauvages cueillirent dans les forêts quelques fruits
nourriciers, quelques racines salutaires, et subvin-
rent ainsi à leurs plus pressans besoins : les premiers pâtres s'aperçurent que les astres suivent
une marche reglée, et s'en servirent pour diriger
leurs courses à travers les plaines du désert.
Telle fut l'origine des sciences mathématiques, et
celle des sciences physiques.

" Une fois assuré qu'il pouvait combattre la
nature par elle-même, le génie ne se reposa plus;
il l'épia sans relâche, sans cesse il fit sur elle
de nouvelles conquêtes, toutes marquées par
quelque amélioration dans l'état des peuples.
Se succédant dès-lors, sans interruption, des
esprits méditatifs, dépositaires fidèles des doctrines acquises, constamment occupés de les
lier, de les vivifier, les unes par les autres, nous
ont conduits, en moins de quarante siècles, des
premiers essais de ces observateurs agrestes aux
profonds calculs des Newton et des Laplace, aux
énumérations savantes des Linnæus et des Jussieu. Ce précieux héritage, toujours accru,
porté de la Chaldée en Egypte, de l'Egypte
dans la Grèce, caché pendant des siècles de malheur et de ténèbres, recouvré à des époques
plus heureuses, inégalement répandu parmi les

peuples de l'Europe, a été suivi partout de la richesse et du pouvoir ; les nations qui l'ont recueilli sont devenues les maîtresses du monde ; celles qui l'ont négligé sont tombées dans la faiblesse et dans l'obscurité."

The first éloge was read on the 5th of April, 1800, and is that of the venerable Daubenton, who, it will be recollected, was the colleague of M. de Buffon, born in the same town with him, and chosen by him to aid his scientific labours. The reasons of this choice are given by M. Cuvier, who first describes Buffon as a man of independent fortune, whose personal and mental attractions, and violent thirst for pleasure, seemed to cast his destiny in any other mould than that of science, but who nevertheless found himself irresistibly drawn towards it, the surest sign of his extraordinary talents. Long uncertain to what object he should devote his genius, he tried several pursuits, and at length fixed on natural history. From the first he measured it in its whole extent ; he, at one glance, perceived what he had to do ; what was in his own power to effect, and in what he required assistance. I would fain quote all that M. Cuvier says of his predecessor ; but a few of the leading points of the different éloges are all that can be offered here, in order to give an idea of their nature, their variety, and their beauty. Continuing to speak of M. de Buffon, M. Cuvier states, that, gifted with the most ardent imagination, and possessing a pen that was the echo of that imagination, viewing nature in all its activity and freshness, and deeply impressed with it as a whole system of beauty and order, he required some one to inspect the details, some one who was gifted with the power of patient investigation, some one whose love of justice and calm tone of mind would form a sort of counter-balance to his ardour, some one equally devoted to the cause, but at the same time modest enough to play a secondary part, and leave him in possession of the brilliant fame he coveted. These requisites were all centred in Daubenton, the companion of his youth. Both morally and physically there was the strongest contrast between the two friends, and each was possessed of those qualities which were necessary to moderate and improve the other. Buffon, commanding every thing, eager for immediate results, and imperious by nature, was desirous of divining the truth, not of reaching it by patient investigation. His imagination at every instant placed itself between him and nature, and his

eloquence seemed to wrestle with his reason
before he employed it in captivating others.
Daubenton, delicate in constitution, moderate
by nature as well as reason, pursued his researches with the most scrupulous circumspec-
tion; he only believed and affirmed that which
he had seen and touched; and far, very far,
from wishing to persuade by other means than
facts, he carefully avoided, both in his writings
and discourse, every figure of speech, and every
fascinating expression. Unalterable in patience,
he was never annoyed at delay; he recom-
menced the same labour over and over again,
until he had succeeded to his satisfaction; and
the method of his proceedings, while it seemed
to call into use every mental resource, seemed to
impose silence on his imagination. When Buffon
first attached him to the Jardin des Plantes, he
thought he had found a laborious aid, who would
smooth the ruggedness of his path; but he found
much more, for Daubenton was a faithful guide,
who pointed out to him the hidden dangers and
precipices of that path. Many times did the sly
smile of Daubenton, when he conceived a doubt,
induce Buffon to reconsider his ideas. Many
times did one of those words, which this friend
knew so well where to place, stop him in his

precipitous career; and the wisdom and prudent
reserve of the one, uniting themselves to the
force and rapidity of the other, gave to the
" Histoire des Quadrupèdes," the only work
common to both, that perfection which makes it
the most interesting part of the great Natural
History of Buffon.   It is more exempt from
errors than the rest, and will long remain a
classical book among naturalists.   Daubenton
was appointed " Demonstrateur du Cabinet
d'Histoire Naturelle," and his salary was gradually augmented from five hundred francs to
four thousand; he was lodged at Buffon's, and
nothing was neglected which could ensure him
that ease of circumstances which is necessary to
every man of letters, every savant who would be
wholly devoted to science.   Daubenton, on his
side, amply repaid these kindnesses by unremitting obedience to the views of his benefactor,
and, at the same time, erected a monument to
his own glory.   Before the time of Daubenton,
the Museum of Natural History was a mere cabinet, and, strictly speaking, only contained the
shells collected by Tournefort for the amusement of Louis XV., when young.   In a very
few years, the whole face was changed.   Minerals, fruits, woods, and shells were brought

from every quarter, and exposed in the most beautiful order; means were taken for discovering the best modes of preserving different parts of organised beings; and the inanimate remains of birds and quadrupeds re-assumed the appearance of life, presenting the slightest details of character to the attentive observer, while they astonished the curious by the variety of their forms and the brilliancy of their colours. Daubenton conceived a vast plan, and, supported by Buffon, profited by the means his credit afforded. No production of nature was excluded from this temple, and a number of anatomical preparations were collected, which, though less agreeable to the eye, were not less useful to the person who did not limit his researches to the exterior of created beings; who endeavoured to make a philosophical science of natural history, and to force it to explain its own phenomena. The study and arrangement of these objects became a real passion for Daubenton; he shut himself up for whole days in the Museum; he arranged the objects in a thousand different ways; he scrupulously examined all their parts; and he tried every possible arrangement until he found that which neither offended the eye nor natural affinities. Thus it is principally to Daubenton that France owes the magnificent museum of the Jardin des Plantes, where we must be struck with the unwearied patience of the man who amassed all these treasures, named them, classed them, displayed their affinities, described their parts, and explained their properties.* A monument equally glorious to the memory of Daubenton is the complete description of this museum, though circumstances prevented him from carrying it farther than the quadrupeds. Reaumur, who had till then swayed the sceptre of natural history, and whose " Memoirs on Insects" are clear, elegant, and highly interesting, jealous of the increasing fame of the two great naturalists, not only attacked Buffon but his friend, whom he considered as the solid supporter of his brilliant rival. Quarrels even took place

* It is impossible to read these pages without being impressed with the application of several of the passages to the author himself, who appears, however, to be perfectly unconscious of the resemblance. At the time he wrote this concerning Daubenton, he was walking with rapid strides in his steps, and how he surpassed him is best told by the state of the whole of the above establishment at the time of M. Cuvier's death. I understand that considerable difficulty has been felt more than once in writing the éloge of M. Cuvier. A selection from his own concerning others might be made with the strictest justice, and the utmost aptitude; and the candid praise he delighted to bestow on his colleagues would thus in every respect be his best eulogium.

in the Academy, and M. de Buffon was obliged
to tax the good offices of Madame de Pompadour,
in order to preserve Daubenton in the rank
which was due to his labours. At length the
insinuations of their enemies seemed to take
effect, and even Buffon began to think, that it
would be more advantageous for himself to publish his " Histoire Naturelle," in thirteen volumes duodecimo, taking away not only the
anatomical parts but the external descriptions;
and he also determined to appear alone before
the public when treating of birds and minerals.
To act thus was not only to wound Daubenton's
feelings, but to injure him in a pecuniary sense.
He might, with reason, have pleaded that it was
an enterprise common to both; but had he asserted his right, he must have quarrelled with
the director of the Jardin ; he must have quitted
the scenes he had, as it were, created, and
which were inseparable from his existence. He
therefore passed over the loss and the affront,
and continued his labours, in a measure consoled
by the regret expressed by all naturalists, when
they saw the History of Birds appear without his
exact descriptions. It is worthy of mention,
that to such a degree did he carry his spirit of
forgiveness, that he afterwards contributed some

N

parts to the " Histoire Naturelle," although his name was never again attached to the work. His intimacy with his friend was also renewed, and continued unbroken till the death of Buffon.

The efforts of Daubenton were far from being confined to the above-mentioned pursuits, and one of the other objects of his endeavours was an attempt to improve the wool of France,|by which means he obtained a popularity which was very useful to him before the Assembly of the Sans Culottes. A certificate of civism was necessary for his personal safety at that stormy period, to obtain which, his titles of Professor and Academician were of little avail; he was at length presented under the title of Shepherd, and in this character he protected the savant. The curious document of this transaction is still in existence.

In 1773, M. Daubenton obtained permission for one of the professorships of the Collége de France to be changed into a chair of Natural History, and also that lectures should be given at the Museum. It was an affecting sight to behold this old man encircled by his disciples, who received his words with a religious attention, a veneration which converted them into so many oracles; to hear his weak and trembling

voice gradually assume its wonted force and
energy, when he tried to inculcate some of
those great principles to which his meditations had given birth, or to develope some useful and important truth. He forgot his years
and his weakness when he could be useful to
young people, or when he performed his duties.
When made a senator, one of his colleagues
offered to help him, by giving lectures for him.
" My friend," he answered, " I cannot be better
replaced than by you, and when age forces me
to resign my duties, be sure that I shall burthen
you with them;" he was then eighty-three.
When thus appointed, he tried to fill his new
station as he had done all others; but in order
to do this he was obliged to change his manner
of living, the regularity of which had, perhaps,
contributed to its long continuance. The season
was very severe; and the first time he assisted
at the meetings of the body to which he was just
elected, he was struck with apoplexy, and fell
senseless into the arms of his colleagues. The
promptest aid could only restore him to life for a
few minutes, during which he evinced that desire calmly to watch the operations of nature
which had hitherto marked his character. He
touched the different parts of his body which

were affected, pointed out the progress of the
paralysis to his attendants, and expired at the
age of eighty-four, without suffering ; so that it
may be said of him, that he attained, if not the
most brilliant, the most perfect happiness for
which man is permitted to hope.

Although confining myself to the principal
features of the above éloge, I have dwelt on it
much longer than will be advisable for the
others.   Two reasons have induced this ; and
the first is, the circumstance of its being one of
the earliest of M. Cuvier's productions which
was read in presence of the Emperor, on whom
it made a great impression.   The natural style
in which it was written, the natural tone in
which it was read, amid the reigning affectation,
produced the happiest effect ; and it was of this
that M. D——, celebrated for his apt remarks,
observed, " At last we have a secretary who
knows how to read and write."   The second
reason is, that it may be offered as a proof of
the innate excellence of M. Cuvier's judgment ;
it is not the work of a man whose reason was ma-
tured by long years of study, whose feelings
have been rendered impartial by age ; but it was
written when the fire of youth is generally apt
to be dazzled by some favourite opinion, is de-

sirous of pointing out its own powers of discrimination by dwelling on the imperfections of
others, and when (fame being then dearest) it is
but too much inclined to steal into its composi-
tions somewhat of self, some allusion to its own
labours and feelings. None of this is perceptible
in the éloge of Daubenton, any more than in the
rest of M. Cuvier's biographical notices : there
is the desire to do honour to his predecessors ;
there we have laid before us the influence that
past labours are likely to shed over the future ;
there is the strict love of justice, pointing out
errors to serve as beacons for those who follow
the same career ; there is the gentle and unwilling exposure of faults, that desire to admit every
circumstance which could palliate the defect ;
there is the benevolent heart that is so evidently
gratified when opportunity is given for commendation ; and in each, and in all together,
we trace the just celebrity which France has attained from her biographical writers.

Although a shorter notice will suffice for the
other éloges, it will be necessary to mention
them all, in order to show the variety of the
subject, and occasionally to introduce an original
passage, not as a better specimen of style than
could be found elsewhere, but as combining

beauty with general interest. M. Lemonnier, the subject of the second, was head physician to Louis XVI., and a botanist; he spent the greater part of his life in trying to introduce useful plants and trees into France; he solaced the poor, and received no reward from them;. he courageously visited his unfortunate master when in prison, and, at eighty-two years of age, died at the herb shop which he had established in order to obtain a livelihood, but where he had been watched over by his nieces with the most devoted attachment, and visited by his friends, who thought his old age rendered doubly honourable by this independent mode of existence.

M. l'Héritier was also a botanist, but of another description, being a strict follower of the system and nomenclature of Linnæus. A curious anecdote, related in this éloge, forcibly developes the character of the man, and at the same time shows the relation he had with England. Always seeking after fresh acquisitions in his favourite science, and delighting in a knowledge of foreign plants, he heard that Dombey had returned from Peru and Chili with an immense collection, for the publication of which he had long sought the necessary funds. L'Héritier obtained the herbarium from Dombey, allowed him an annual pension, and from that moment no bounds were set to his zeal ; painters and engravers were employed, and the work was far advanced, when he received intelligence, that the Spaniards who had accompanied Dombey demanded of the French government that his botany should not be published before theirs, and, consequently, that the herbarium should be restored to Dombey. The order for this restoration was expected the next day, when L'Héritier, consulting only his friend, M. Broussonet, sent for twenty or thirty packers, and the night was passed in making cases. L'Héritier, his wife, and MM. Broussonet and Redouté, packed the herbarium : early the next morning the former posted off to Calais with his treasure, nor rested till he was safe on the English soil. He passed fifteen months there in the most perfect retirement, and was delighted with the kindness he received. The library and collections of Sir Joseph Banks, the herbarium of Linnæus, then in the possession of Sir J. E. Smith, besides the acquisitions of other botanists, were all open to him, and he there finished his manuscript. The plates were most of them completed when he returned to France ; but political circum-

stances, and the duties he was called on to perform as a citizen, prevented the appearance of
this great work. The same zeal and activity,
united to a most conscientious fulfilment of the
labours allotted to him, distinguished him as a
magistrate; but neither public nor private virtues could save him from the hand of the as-
sassin. Returning home late one evening from
the Institute, he received several stabs from
a sword, and was found dead, the next morning, a few paces from his own door.

M. Gilbert was chiefly celebrated as an agriculturist; and he it was who was sent to Spain
by the government of France, to procure those
beautiful breeds of sheep from that country
which had caused such improvements in the
English wool. This excellent man's character
may be comprehended, when it is known that a
friend of his being suspected, and consequently
imprisoned, during the revolution of 1793, he
every month carried to the wife of the sufferer
the half of his own income, leading her to suppose that the money came from her husband, in
order to prevent her from being aware of the
destitute state into which she was plunged, or
the danger incurred by one so dear. Full of
hope, M. Gilbert started on his mission to Spain

with the most enthusiastic pleasure, little fore-
seeing the obstacles and difficulties he should
encounter. Badly supported by his government, at times wholly neglected, he for two
years was unable even to make the proper pur-
chases, and at length was obliged to pledge his
own property in order to extricate himself from
the embarrassments caused by the conduct of
those in whose promises he had confided. He
had flattered himself that all would have been
completed in three months, but after two years
of painful travelling, incredible fatigue, opposition, and even humiliation of every kind, the
flock he had assembled was scarcely, by one
third, equal to what it ought to have been. His
bodily strength at last yielded to all these sufferings, and he was carried off by a malignant
fever, after an illness of nine days.

Darcet, the confidential friend of Montes-
quieu, his assistant in collecting and arranging
the immense materials for the " Esprit des
Loix," and the preceptor of the young Montes-
quieu, never lost sight of his chemical researches
amid these duties, and he discovered and caused
the execution of wonderful improvements in the
porcelain of France.

The history of Dr. Priestley is too well known

to need much detail here ; but as it is one of the most beautiful pieces of biography which has emanated from the pen of M. Cuvier, I shall cite a passage, in his own words, concerning the labours of this great chemist and natural philosopher. " Priestley, comblé de gloire, s'étonnait modestement de son bonheur, et de cette multitude de beaux faits que la nature semblait n'avoir voulu révéler qu'à lui seul. Il oubliait que ses faveurs n'étaient pas gratuites, et que si elle s'était si bien expliquée, c'est qu'il avait su l'y contraindre par une persévérance infatigable à l'interroger, et par mille moyens ingénieux de lui arracher des réponses.

" Les autres cachent soigneusement ce qu'ils doivent au hasard ; Priestley semble vouloir lui tout accorder : il remarque, avec une candeur unique, combien de fois il en fut servi sans s'en apercevoir, combien de fois il posséda des substances nouvelles sans les distinguer ; et jamais il ne dissimule les vues erronées qui le dirigèrent quelquefois, et dont il ne fut désabusé que par l'expérience. Ces aveux firent l'honneur à sa modestie sans désarmer la jalousie. Ceux à qui leurs vues et leurs méthodes n'avaient jamais rien fait découvrir, l'appelaient un simple faiseur des expériences, sans méthode et sans vues : ' il n'est

pas étonnant,' ajoutaient-ils, 'que, dans tant d'essais et de combinaisons, il s'en trouve quelques-
uns d'heureux.' Mais les véritables physiciens
ne furent point dupes de ces critiques inté-
ressées." *

There is yet another passage which, while it
so ably pleads the cause of Priestley, places
M. Cuvier's candour in so conspicuous a light,

---

* Priestley, loaded with glory, was modest enough to be
astonished at his good fortune, and at the multitude of beautiful facts which nature seemed to have revealed to him alone.
He forgot that her favours were not gratuitous, and if she
had so well explained herself, it was because he had known
how to oblige her to do so by his indefatigable perseverance
in questioning her, and by the thousand ingenious means he
had taken to snatch her answers from her.

Others carefully hide that which they owe to chance;
Priestley seemed to wish to ascribe all his merit to fortuitous
circumstances, remarking, with unexampled candour, how
many times he had profited by them without knowing it, how
many times he was in possession of new substances without
having perceived them; and he never dissimulated the erroneous views which sometimes directed his efforts, and from
which he was only undeceived by experience. These confes-
sions did honour to his modesty, without disarming jealousy.
Those to whom their own ways and methods had never discovered anything, called him a simple worker of experiments,
without method and without an object; "it is not astonishing," they added, "that among so many trials and combin-
ations, he should find some that were fortunate." But real
natural philosophers were not duped by these selfish criti-
cisms.

that I shall make no apology for introducing it,
though it will not be necessary to give it in
French. " I am now, Messieurs, arrived at the
most painful part of my task. You have just
seen Priestley successfully progressing in the
study of human science, to which he nevertheless consecrated but a few of his leisure moments. I must now present him to you in
another light, wrestling against the nature of
those things which are hidden from our reason
by an impenetrable veil, trying to submit the
world to his conjectures, consuming almost all
his life in these vain efforts, and at length plung-
ing himself into an abyss of misery. Here, like
himself, I have need of all your indulgence;
perhaps the details into which I am about to
enter will, to some, appear foreign to the place
in which I speak, but it is here, I think, that
the terrible example they give ought to be heard
with the greatest interest. I have already told
you that Priestley was a minister of religion, and
I am forced to add, that he professed four different creeds before he could decide on teaching
one of them in his public capacity. Brought up
in all the severity of the presbyterian faith,
which we call Calvinistic, and in all the bitter-

ness of predestination, such as Gomar taught it,
he scarcely began to reflect, before he turned to
the milder doctrine of Arminius. But, as he
advanced, he always seemed to find too much to
believe; he therefore adopted the tenets of the
Arrians, who, after having invaded Christianity
from the time of the successors of Constantine,
have now no other asylum than in England, but
whose faith is decorated by the names of Mil-
ton, Clarke, and Locke, and even, as report
says, that of Newton, and whose reputations, in
some measure, repair the loss of former power.

" Arrianism, while it declares Christ to be a
creature, believes him, nevertheless, to be a
being of a superior nature, produced before the
world, and the organ of the Creator in the production of other beings. This is the doctrine
clothed in the magnificent poetry of the Paradise
Lost. After having long professed this, Priestley
abandoned it, in order to become an Unitarian,
or that which we call Socinian. There are few,
perhaps, among those who now hear me, who
have ever informed themselves in what these
two sects differ. It is, that the Socinians deny
the pre-existence of Christ, and only look upon
him as a man, though they revere in him the

Saviour of the world; and they acknowledge that the Divinity was united to him, in order to effect this great work. This subtle shade between two heresies, for thirty years occupied that head which was required for the most important questions of science, and, without comparison, caused Priestley to write more volumes than he ever produced on the different species of air . . . . . . His last moments were full of those feelings of piety which had animated his whole life, the improper control of which had been the foundation of all his errors. He caused the Gospel to be read to him, and thanked God for having allowed him to lead an useful life, and granted him a peaceful death. Among the list of his principal blessings, he ranked that of having personally known almost all his contemporaries. ' I am going to sleep, as you do,' said he to his grandchildren, who were brought to him, ' but we shall wake again together, and, I hope, to eternal happiness; ' thus evincing in what belief he died. These were his last words; such was the end of that man, whom his enemies accused of wishing to overthrow all morality and religion, and, nevertheless, whose greatest error was to mistake his vocation, and to attach too much importance to his individual sentiments, in matters where the most important of all feelings ought to be the love of peace."

The subject of the succeeding éloge, M. Cels, was a practical botanist and scientific agriculturist, to whom Paris owes the celebrated garden which bears his name : from him emanated some excellent laws on agricultural interests.

No one but a profound naturalist could have appreciated the merits of M. Adanson ; and no one but an impartial and penetrating biographer could have separated his great and rare perfections, from that peculiarity and exaggeration of ideas which led him into error. This traveller visited Senegal, because it is the most difficult of access, the most unhealthy, and, in all respects, the most dangerous of all the French colonies, and, consequently, was the least known to naturalists ; the continent of Africa was therefore the scene of his discoveries, and to him we owe our perfect knowledge of that giant of the vegetable world, the Baöbab, or, in proper terms, the Adansonia digitata.

M. Broussonet, Professor of Botany to the School of Medicine at Montpelier, was called to the Institute by the section of zoology and anatomy, and after publishing several works on zoology, and passing a life of dangers and unheard-of escapes, died of a coup de soleil.

M. Lassus was a surgeon, and though generally skilful in his profession, was so unfortunate as to bleed a royal patient twice without success. The outcry was universal. " Une princesse piquée deux fois, et qui n'a pas saigné — quel accident effroyable !" said the courtiers ; the physicians shook their heads with a mysterious look ; but the princess, being more generous, procured M. Lassus a situation in place of that from which she had been obliged to dismiss him in her household, and by so doing, secured a meritorious and devoted servant, both to herself and the public. With her and her sister he travelled over Italy, at the time of the great revolution ; and by producing his portfolios as proofs that he had enriched his country with useful information, evaded the law against emigrants, which would have been enforced against him on his return, and was appointed to the medical school at Paris.

M. Ventenat was a priest and botanist, and, protected by Josephine, described the treasures of her garden at Mal Maison.

The name of De Saussure will ever be dear
to geologists; and with his éloge, and that of his
uncle, M. Bonnet, the naturalist of Geneva, the
first volume closes. In this combined éloge is
a passage in which M. Cuvier's talents for description show themselves; and as it is almost
an isolated instance in his published writings, I
here quote it : — " Comme le voyageur est ravi
d'admiration, lorsque, dans un beau jour d'été,
après avoir péniblement traversé les sommets du
Jura, il arrive à cette gorge, où se deploie su-
bitement devant lui l'immense bassin de Genève,
qu'il voit d'un coup d'œil ce beau lac dont les
eaux réfléchissent le bleu du ciel, mais plus pur
et plus profond; cette vaste campagne, si bien
cultivée, peuplée d'habitations si riantes; ces cô-
teaux qui s'élèvent par degrès et que revèt une
si riche végétation, ces montagnes couvertes de
forèts toujours vertes; la crête sourcilleuse des
Hautes Alpes, ceignant ce superbe amphithéâtre,
et le Mont Blanc, ce géant des montagnes Eu-
ropéennes, le couronnant de cette immense
groupe de neiges, où la disposition des masses et
l'opposition des lumières et des ombres, pro-
duisent un effet qu'aucune expression ne peut
faire concevoir à celui qui ne l'a pas vu.

" Et ce beau pays, si propre à frapper l'imagination, à nourrir le talent du poëte où de l'ar-
tiste, l'est, peut-être, encore davantage à reveiller
la curiosité du philosophe, à exciter les recherches du physicien. C'est vraiment là que la
nature semble vouloir se montrer par un plus
grand nombre de faces.

" Les plantes les plus rares, depuis celles des
pays tempérés jusqu' à celles de la Zone Glaciale,
n'y coûtent que quelques pas au botaniste ; le
zoologiste peut y poursuivre des insectes aussi
variés que la végétation qui les nourrit ; le lac
y forme pour le physicien une sorte de mer, par
sa profondeur, par son étendue et même par la
violence de ses mouvemens ; le géologiste, qui
ne voit ailleurs que l'écorce extérieure du globe,
en trouve là les masses centrales, relevées et per-
çant de toute part leurs enveloppes, pour se
montrer à ses yeux ; en fin, le météorologiste y
peut à chaque instant observer la formation des
nuages, pénétrer dans leur intérieur, ou s'éléver
au-dessus d'eux." *

* How delighted is the traveller when, in a beautiful
summer's day, after having with difficulty traversed the sum-
mits of the Jura, he arrives in this ravine, where the immense
basin of Geneva suddenly opens before him, when at one
glance he sees this beautiful lake, the waters of which reflect

The second volume opens with the éloge of Fourcroy, — the brilliant, the eloquent, the calumniated Fourcroy. The struggles of his youth, and his vigorous resistance of injustice and poverty, the account of his discoveries, — all form one of the most powerful pieces of biography

---

the blue of heaven more deeply and more purely ; this vast country, so well cultivated, and peopled by smiling habitations; the hills, which rise by degrees, clothed with the richest vegetation ; the mountains, covered with evergreen forests; the frowning crests of the High Alps, above this superb amphitheatre ; and Mont Blanc, the giant of European mountains, crowning the immense group of snows, where the disposal of the masses, and the contrasts of light and shade, produce an effect which no expression can convey to those who have not seen it.

And this beautiful country, so calculated to strike the imagination, to feed the talent of the poet or the artist, is perhaps still more so to awaken the curiosity of the philosopher, and to excite the researches of the follower of natural philosophy. It is truly there that nature seems to delight in showing herself under a number of different aspects.

The rarest plants, from those of temperate countries to those of the Frozen Zone, only cost the botanist a few steps. The zoologist may there pursue insects as varied as the vegetation which nourishes them. The lake there forms, from its depth and extent, and even its violent movements, a sort of sea for the natural philosopher; the geologist, who, elsewhere, sees but the external rind of the globe, there finds central masses, thrown up, and in every part piercing their envelopes, and showing themselves to his eyes; lastly, the meteorologist can there observe the clouds at every instant, penetrate within them, or raise himself above them.

ever read, The following description of his
lectures recalls those of the author, and, in many
instances, is equally applicable to both : —" For
five and twenty years the amphitheatre of the Jardin des Plantes was the centre of M. Fourcroy's
glory. The great scientific establishments of
this capital, where celebrated masters expose to
a numerous public, capable of passing judgment
on them, the most profound doctrines of modern
times, recall to our memory that which was
noblest in antiquity. We fancy we again find
in these assemblies a whole people animated by
the voice of a single orator; and again see those
schools, where chosen disciples came to penetrate the oracles of a sage. The lectures of
M. Fourcroy corresponded to this twofold picture : Plato and Demosthenes seemed to be
united in him ; and it is almost necessary to be
one or the other, to give an idea of them. Connection of method, abundance of elocution, elevation, precision, elegance of terms, as if they
had been selected long beforehand; rapidity,
brilliancy, novelty, as if suddenly inspired; a
flexible, sonorous, and silvery voice, yielding to
every motion, penetrating into the corners of the
largest audience-room ; — nature had bestowed
every thing on him. Sometimes his discourse

flowed smoothly and majestically; the grandeur
of his metaphors, and the pomp of his style, were
all imposing; then, varying his accents, he
passed insensibly to the most ingenuous fami-
liarity, and fixed attention by sallies of the most
fascinating gaiety. Hundreds of auditors, of all
classes, all nations, were to be seen, passing
whole hours, closely pressed against each other,
almost fearing to breathe, their eyes fixed on
his, suspended to his mouth, as the poet says
(pendent ab ore loquentis). His look of fire
darted over the crowd; in the farthest rows he
distinguished that mind which was difficult to
convince, and still doubted, or the slow comprehension which did not completely understand;
for these he redoubled his arguments and his
similes, and varied his expressions until he
found those which would convince; language
seemed to multiply its riches for him, and he did
not quit his subject till he saw all his numerous
audience equally satisfied."

It is scarcely possible to mention Fourcroy,
without recollecting the odious suspicion attached
to his name*; I therefore give M. Cuvier's observ-

* It was reported that he might have saved the life of
M. Lavoisier during the reign of terror, as indeed he had saved
many by his influence; but, at the moment of M. Lavoisier's

ations, taken from the same éloge : — " Perhaps I may be blamed for recalling these sad recollections; but where a celebrated man has been so unfortunate as to be accused, as M. Fourcroy was, — where this accusation occasioned the torment of his life, — the historian would in vain strive to bury it in oblivion, by being himself silent. We ought now to say, that if, in the strict researches we have made, we had found the slightest proof of so horrible an atrocity, no human power could have forced us to sully our lips by his éloge, to make the roofs of this temple resound with our praises,— this temple, which ought to be no less the asylum of honour than of genius."

To Dessesserts, the physician, and subject of the next éloge, the French owe the banishment of those horrible machines of whalebone, those swathing clothes, those hot-houses, where the minds and bodies of infants were imprisoned from their birth. By M. Dessesserts were those mothers recalled to their duty, who abandoned

---

arrest, his own life was threatened, and all power of being useful to others was taken from him. Lavoisier fell a victim to the revolutionary monsters, and M. Fourcroy was accused of taking a part in that which freed him from a powerful rival.

the nourishment of their offspring to others,
when capable of affording it themselves; and,
though unacknowledged, to M. Dessesserts was
Rousseau indebted for the first pages of his
Emile.

The next subject of biographical notice is
Henry Cavendish, that remarkable Englishman,
who, notwithstanding his splendid fortune and
his noble birth, pursued science with the most
disinterested ardour. How M. Cuvier appreciated his labours, will be gathered from the following passage : — " All that science revealed
to him, seemed to be tinctured with the sublime
and the marvellous : he weighed the earth, he
prepared the means of navigating through the
air, he deprived water of its elementary quality ;
and these doctrines, so new, and so opposed to
received opinions, were demonstrated by him in
a manner still more extraordinary than the discovery itself. The writings where he lays them
before others, are so many chefs d'œuvre of wisdom and method; perfect in their whole, and
perfect in their details, in which no other hand
has found any thing to reform, and the splendour
of which has only increased with time . . . .
so that there can be no temerity in predicting,
that he will reflect back upon his house much

greater lustre than he has received from it; and
that these researches, which, perhaps, excited
the pity and contempt of some of his contemporaries, will make his name resound, at an age to
which his rank and his ancestry alone would not
have transmitted it. The history of thirty centuries clearly teaches us, that great and useful
truths are the sole durable inheritance which
man can leave behind him."

The next in the list of great names is that of
Pallas, the enlightened and sagacious traveller
of the north of Asia, the inhabitant of the
Crimea, and the learned and indefatigable naturalist.

The éloges of M. Parmentier and Count Rumford are combined, and commence with a sort of
introduction to the useful labours of each; labours which bore so strongly on the means of
affording warmth and nourishment to the poorer
classes. The former, who had learned the value
of the potato as an article of food in the prisons
of Germany, overcame the prejudices entertained
against them in France, where they were said to
produce leprosy, fevers, and no one knows what
diseases. His mode of rendering them popular
and desirable was curious; for he began by cultivating them in the open fields, and causing

them to be carefully guarded by day only : he was but too happy when he was informed, that this apparent caution had induced depredation by night. He then obtained from the king of France the favour of wearing a bunch of potato blossoms in the button-hole of his coat, at a solemn fête; and nothing more was required to cause some of the great lords of the kingdom to order its cultivation on their estates. Not, however, till the last years of his life, was he completely successful; and during the great Revolution he was rejected as a magistrate, because he had *invented* potatoes.

Benjamin Thomson, Count Rumford, was an American by birth, and served as a royalist in the war between America and England. After the peace he came to the latter country, where he was knighted by George III., and recommended by that sovereign to the protection of the Elector of Bavaria, at whose court he rose to the highest dignities. It was then that he turned his attention to the state of the poor, and, in trying to find means for ameliorating their condition, he made those beautiful discoveries which have benefited all classes.

The labours and character of the oriental traveller, Olivier, are then noticed, and the history of this excellent man furnishes another proof of the immense influence, that a knowledge of medicine will produce among uncivilised people.

M. Tenon, the surgeon, is afterwards presented to us. His youth was passed in a series of struggles; his maturity was beautiful, and he reached the age of ninety-two without intellectual infirmity.

The éloge of the famous Werner is in every respect interesting, for in it we find a brief résumé of all that was done by this great man, together with the peculiarities which deprived the world of the written results of his labours and extensive knowledge; he having preferred to trust his reputation to the justice of his disciples, rather than have recourse to his own pen for transmitting it to posterity.

The life of Desmarets follows; — Desmarets, the antagonist of Werner, the champion of volcanoes; he in whose discoveries originated the famous disputes between the Plutonians and Neptunians, and which disputes not only placed the whole world between fire and water, but occasioned more animosity than any question which had hitherto agitated the learned world.

To this second volume are added two éloges read before the Philomathic Society of Paris, the

discourse of M. Cuvier on his reception at the
Académie Française, and the reply of the director of that academy. The first of these two
éloges is that of M. Riche, whose life resembles
that of a hero of romance, and whose feelings
and adventures, perhaps, caused his death at
the age of thirty-five. The second is that of
M. Bruguière, the companion of Olivier, already
noticed. The discourse of M. Cuvier assumes a
tone in which the nature of his professional
studies scarcely ever allowed him to indulge,
but in which we trace the same perfection as
elsewhere. It is full of classical and elegant
allusions; it is the production of a man of
letters, and shows how admirable is the combination when science and literature occupy the
same mind. In the reply of the Count de Sèze
will be found a very admirable résumé of M.
Cuvier's labours up to that period.

The third volume begins with the éloge of
M. de Beauvois, the African traveller, to whom
the world owes the Flora of Owaree and Benin;
and who, after wrestling with the storms both of
this continent and those of America, died in
consequence of the sudden changes to which an
European climate is so frequently liable. In

this biography are some remarkable passages
concerning slavery.

M. Cuvier's brotherly feeling, — his gratitude, if
I may so express myself, — towards all promoters
of science, is nowhere more strongly manifested
than in his eulogium on Sir Joseph Banks, the
distinguished and munificent patron of scientific
labourers. The travels and adventures of Sir
Joseph are here related with vivacity; and the
famous dispute about points and buttons to elec-
trical conductors, which placed him at the head
of the Royal Society, and which, in other hands,
might have afforded much scope for ridicule, is
touched on with a delicacy peculiar to M. Cuvier's disposition. Nor is this éloge less remarkable for the honourable testimony given to a
nation which has been but too often regarded
with jealousy, and which has but too often met
these sentiments with a reciprocal feeling. " The
savans of England," says the Baron Cuvier,
" have taken an equally glorious part in those
mental labours which are common to all civilised
people: they have confronted the eternal frosts
of either pole; they have not left a corner of the
two oceans unvisited; they have increased the
catalogue of nature tenfold; heaven has been
peopled by them with planets, satellites, and un-

heard-of phenomena; we may almost say that
they have counted the stars of the milky way.
If chemistry has assumed a new aspect, the
facts they have furnished have essentially contributed to this metamorphosis. Inflammable
air, pure air, phlogistic air, are due to them;
they have discovered the decomposition of water,
and a number of new metals have been produced
by their analyses. The nature of fixed alkalies
has only been demonstrated by them; mechanism, at their voice, has given birth to miracles,
and placed their country above all others in
almost every species of manufacture."

The mineralogist, M. Duhamel, appeared at a
time when De Saussure had not travelled, Deluc
had not written, nor Werner, by the force of his
extraordinary genius, arranged the mineral universe; and, after years of scientific labour, was
appointed to the Ecole des Mines, established in
Paris; and in tracing his influence in this professorship, M. Cuvier thus speaks : —" Our pro-
ducts in iron are quadrupled; the mines of this
metal opened, near the Loire, in the region of
coal, and in the midst of combustible matter, are
about to yield iron at the same price as in England. Antimony, manganese, which we formerly imported, are now exported in considerable

quantities. Chrome, discovered by one of our chemists, is also the useful product of one of our mines. Zinc and tin have already been extracted from the mines on the coast of Britany. Alum and vitriol, formerly almost unknown in France, are collected in abundance. An immense mass of rock salt has just been discovered in Lorraine; and all promises that these new creations will not stop here. Doubtless, it is not to a single man, nor to the appointment of a single professorship, that all this may be attributed ; but it is not the less true, that this one man, this one professorship, has been the primary cause of these advantages."

The name of M. Haüy, the geologist, the mineralogist, the founder of crystallography, forms a sort of oracle in the learned world, and I have a peculiar pleasure in dwelling on this éloge, because it is one of the most admirable of all, and does honour to M. Cuvier's heart, showing how entirely he was independent of selfish feelings, how truly just he could be, even to those who had opposed him with hostile sentiments. The extraordinary man here spoken of commenced the world as a chorister, and studied natural philosophy and botany as amusements. These tastes led him frequently to the Jardin des

Plantes, in Paris; and chance took him one day, with the crowd, into the amphitheatre, to hear M. Daubenton lecture on mineralogy. Mineralogy henceforth became interesting to him; and chance equally befriended him in this new direction of his pursuits. Happening to examine a mineral at the house of a friend, he accidentally let fall a beautiful group of calcareous spar; the fracture of one of the prismatic crystals opened a new world of ideas to him, and he became the M. Haüy, the legislator of mineralogy, the founder of a system which has been adopted all over the world. Imprisoned during the fury of the Revolution, he tranquilly pursued his studies in his cell, and was with difficulty torn from it by his friend, M. Geoffroy St. Hilaire, on the fatal 2d of September. In 1802, he was appointed professor to the Museum of Natural History. Pious, benevolent, tolerant, and devoted to his studies, no worldly considerations ever intercepted his religious exercises nor his scientific labours; and his mode of living was as simple as the station from which he sprung: he walked in the same places every day, took the same exercise, wore the same fashion of clothing, and his manners and language were equally remarkable for their primitive simplicity. A fall

in his own room occasioned a fracture from which
he never recovered; but, during the long hours
of pain which preceded his death, he divided his
time between prayer, a careful edition of his
works, and the future fate of his pupils.

Count Berthollet was a chemist of the most
elevated rank; and to him is due the discovery
of the present method of bleaching linen, and
many improvements in dyeing.

M. Richard came into the world at Auteuil,
a garden belonging to Louis XV., of which he
afterwards became the chief; and, born in the
midst of plants, he knew their names before he
could read, and could draw them before he
could write correctly. To the study of botany
was his whole life devoted; for this he perfected
himself in drawing, and became acquainted with
the Greek and Latin languages; for this he refused advantageous offers in the church; and
for this he was turned out from the paternal
dwelling, with the scantiest pittance. Drawing
by night, and studying botany by day, he by
degrees accumulated money, but this money was
for his favourite science. He was sent to the
French colonies in America, to propagate Indian productions, and discover which of theirs
could in turn be made useful. Laden with

treasures he returned to France, but all there
was changed; M. de Buffon was dead; the
government unskilful and in confusion; no one
recollected the promises made to him, and
people whose heads were hourly in danger,
cared little for the cloves of Cayenne. En-
feebled in health, exhausted in fortune, and unable to look forward to better times, M. Richard had to recommence the same sort of life
which he had led at fourteen years of age. As
a man of science he remained as great as ever;
his dissertations were astonishing proofs of the
extent and sagacity of his views; but his temper,
soured by so many misfortunes, never recovered
its tone, and he died, at the age of sixty-seven,
after much bodily and mental suffering.

Few who have been in the habit of visiting
the Jardin des Plantes within the last forty
years will be ignorant of the name, at least, of
M. Thouin. He there succeeded his father as
head gardener, and uniting science and the
most enlightened views to practical knowledge,
and placing his affections on the improvement
of his garden, he became a centre of correspondence for all parts of the world. His fine
countenance, noble and engaging deportment,
and his interesting conversation, caused him to

P

be sought by the most elevated, as well as the most humble, in the ranks of life. He died in 1824.

The Count de Lacépède is presented to us in three different points of view; first, as a practical and theoretical musician of considerable skill; secondly, as a man of science; and, thirdly, as a statesman; and crowning the whole by mingling the most invariable politeness, the most amiable deportment and feeling, and highest moral excellence, with all his duties. He died, at the age of sixty-nine, of the smallpox.

The éloges of MM. Hallé, Corvisart, and Pinel, three great physicians, are united into one. The first of these was the active promoter of vaccination, was skilful in his treatment of chronic disorders, and was equally celebrated for his charity. M. Corvisart, who lost several opportunities of promotion because he would not wear a bag wig, was at length appointed to the direction of the Hôpital de la Charité, and afterwards to a professorship at the Ecole de Médecine. His fame spread through Europe, and, before he died, he became head physician at court. M. Pinel prepared himself for the study of medicine by a knowledge of mathematics and natural history, but, unable to express himself, in con-

sequence of a most invincible timidity, he was
long neglected. When however, his merits
once became known, he rose rapidly in fame;
he was appointed to the hospital of Bicêtre;
thence to that of the Salpêtrière, and afterwards
to a chair at the Ecole de Médecine. He was
particularly famous for his classification of dis-
eases, and his treatment of madness.

It would be impossible, in the brief sketch to
which I am limited, to do justice to the éloge of
M. Fabbroni, who, from the variety of his genius
and knowledge demanded equal variety from his
biographer; and all that can be done is to show a
portion of the talents which have elicited this
remark. Like most of those who have attained
great celebrity, the early years of M. Fabbroni
were passed in struggle and difficulty. His first
work was entitled, " Reflections on the Present
State of Agriculture; or, An Exposition of
the True Method of cultivating (landed) property." He became sub-director of the beautiful museum at Florence, where he founded
lectures. Driven from this establishment by
Marie Louise, Queen of Etruria, he yet continued
to serve his country; and while carrying on
various administrative duties, published his own
useful ideas concerning the arts, agriculture,

political economy, and the general questions connected with the profoundest theories of science. The wines of Italy were greatly improved by his means; and Tuscany, being destitute of fuel, the grand Duke applied to M. Fabbroni to assist in finding coal; and both Leopold and his son continued to protect him, and to profit by his administrative and scientific talents. When all Italy was alarmed at the conquests of the French, M. Fabbroni sought refuge in his chemical studies, as applied to the useful arts; and when Tuscany had recognised the French republic, he was charged with a mission to France, concerning the unity of weights and measures. Being in Paris at the time that war was declared against Austria and Tuscany, he obtained permission for a special conservator to be sent to Florence to preserve the collections there; and in consequence of his care, the only thing taken from thence was the Venus de Medicis, and which, in fact, had been clandestinely abstracted before the arrival of the French, and given up to them by the King of Naples. The whole of M. Fabbroni's life was a scene of active service; and we find him, at one time, charged with delicate political missions; at others, with the direction and administration of the mint at

Florence ; seeking the causes of pestilence, and
the means of prevention; making roads, fixing
conductors for lightning, and aiding the state by
his counsels. France employed him in the departments beyond the Alps, as director of bridges
and highways; and in this capacity he caused
new roads to be made in every direction, bridges
to be thrown over fearful torrents, and two magnificent military causeways, which, raised along
precipitous crests, supported by arches of prodigious elevation, and occasionally piercing the
bosom of these rugged mountains, have made an
agreeable walk of that which was formerly frightful to the imagination.

To these two éloges succeed two funeral discourses ; one delivered at the interment of M.Van
Spaendonck, the professor of botanical drawings at the Jardin des Plantes, an artist whose
productions attained the highest perfection; and
the other at the grave of the great astronomer,
M. Delambre. The latter was a personal friend
of M. Cuvier's ; and in this discourse, which was
not of sufficient extent to admit of an enumer-
ation of his labours, his excellent character as a
man received its just tribute from the lips of his
colleague.

The volume is closed by two of those admirable reports, in which M. Cuvier always displayed his genius and acquirements in their full strength. In the first, which is on the progress of natural history between the maritime peace and the year 1824, will be found an account of the important travels of that period. The second treats of the principal changes which chemical theories have undergone, and of the new services rendered by this science to society at large, and was read at a general meeting of the four academies, in 1820.

The forthcoming volume of these éloges will, if nothing unforeseen should occur, be shortly published, and will contain those of M. Ramond, the Pyrennean traveller; M. Bosc, the successor of M. Thouin; Sir Humphrey Davy, M. Vauquelin, and M. Lamark; some funeral orations; M. de Lamartine's discourse on his reception as a member of the Institute, with M. Cuvier's reply; and a new edition of the Prix de Vertu. These have all been read in public; but of course, when printed, a freer scope is given to detail; for no one knew better than M. Cuvier how to fascinate a numerous audience, by a choice of what was generally interesting, or to avoid the ennui produced by too long a demand on their attention.

It is for ever to be regretted, that the last course of lectures delivered by M. Cuvier has been comparatively lost to mankind in general. The hall at the Collège de France resounded with these luminous discourses, taken at the moment from mere memoranda, and now only existing in the memory of his auditors. He was extremely averse to short-hand notes, because he thought them very inadequate to the purposes of publication; and he had no time, he said, either to edit them himself, or correct the editions of others. The glimpses (for they can only be called such) given in the feuilletons of the Temps, and in the pamphlets compiled by M. Magdeleine de Saint Agy, were then published entirely without his sanction, and the latter even without his knowledge; but imperfect as they are, they yet assist in giving a general idea of the plan that was followed.

Conscientiously fulfilling some of the most important duties of the state, equally devoted to those of his different secretaryships and professorships, and daily progressing in the most profoundly scientific works and discoveries, it is no wonder that he rarely found time for a course of lectures. At length, however, struck with the errors which he perceived in the system of unity

of composition, and fearing the injurious direction that such ideas might give to youthful studies, he combated them solely for the love of
science; and his health fortunately permitting,
he for this purpose resumed his chair at the
college, and, taking for his subject the entire
history of natural sciences, he, in this series,
seemed to carry learned research, precision,
clearness, sound and elevated views proceeding
from the deepest thought and erudition, and
a pre-eminent power of separating truth from
error, to the highest degree to which man could
attain. The charms of his flexible and sonorous
voice, which could be heard every where in its
sweetest tones, the benignity and animation of
his countenance, attracted each sex and various
ages. In the coldest weather, the audience
assembled an hour before the time, and some
were contented to remain on the staircase, provided they could catch some of his melodious
words; and the enthusiasm with which he was
received, while it endangered his personal convenience, called forth that benevolent smile
which was calculated rather to encourage than
repress these marks of admiration.

" The fundamental principle of these lectures," says M. Laurillard, " was, that society

having been developed by the discovery of the
natural properties of bodies, each of these discoveries has a corresponding degree of civilisation; and therefore the history of this civilisation, and consequently of all humanity, is
intimately connected with the history of natural
sciences." In order to be fully in possession of
his subject, how immense must have been the
research of M. Cuvier! and nothing but a
review of his whole life seems to account for his
capability. Several have been able to elucidate
particular periods, to the study of which they
have devoted themselves; but his researches embraced all historical and philosophical science.
He consulted all books, in order to go back to
the origin of discoveries; and the judgment
necessary for the employment of materials thus
collected was so much the greater, inasmuch as
writers frequently state but the germs of their
ideas, and leave facts almost as obscure as they
are in nature.

The first, or opening lecture, divided the progress of science into three epochs; the religious,
more especially emanating from the Egyptians and
Hebrews; the philosophical, which commenced
in Greece; and the third, the beginning of
which may, perhaps, be traced to Aristotle,

though its importance can only be dated from the sixteenth century. In this lecture were also discussed the age of the world, the vestiges of the great deluge, and the value of the astronomical records of primitive nations.

The second lecture gave a sketch of the four great nations constituted at the remotest period before Christianity, and of which history gives us any certain information. The extent of their knowledge was measured; the influence of that knowledge appreciated; and, in speaking of Moses, M. Cuvier said that, although Moses was brought up in all the learning of the Egyptians, he foresaw the inconveniences of, and laboured much to abolish their practice of veiling the truth under mysterious emblems. That Moses was in possession of that truth was evident from his system of cosmogony, which every discovery of recent times serves but to confirm. The progress of the nations who sprung from the Egyptians, the diffusion of their learning, the bards, the philosophers, the schools of Greece, were given with a most absorbing interest and beauty, and occupied six lectures. In the eighth, he began his history of Aristotle, the founder of the science of natural history. As might be expected, M. Cuvier became, if possible, more

eloquent, more fascinating than ever. The subject was likely to inspire him, and his audience
were not disappointed; they left the lecture-
room,. forgetting their favourite professor, for
the moment, in his description of his great predecessor.

The twelfth lecture was devoted to the advantages which accrued to science, in consequence of the labours of Aristotle. From these
the Professor passed to a rapid sketch of the
history of the Ptolemies; and before he laid the
world before his hearers, in the state in which
it was under the dominion of the Romans, he
glanced over the Carthaginians and Etrurians.
Having at length reached the masters of the
globe, he gave a full description of those magnificent feasts, and those combats of animals,
which put every known quarter of the earth
under contribution, and passed all their learned
men in review. Then tracing the state of science during the great struggles which established
Christianity, and during its languid existence in
the Byzantine Empire, M. Cuvier led the attention towards the Arabs, who cultivated some
branches with success. He then followed it
into the different nations composed of the wrecks
of the Western Empire, and through the slight

glimmerings of existence shown during the
middle ages, and throwing the same deep tone
of interest over every epoch, the revival of letters gave fresh scope to his discourse. It was
no longer a mere dawning, or a decay, which at
times seemed hopeless; but it was a series of
brilliant discoveries, which spread their influence over the remotest parts of the world; and,
beginning with printing, he, in his opening lecture to the second part of his course, premised,
that he should no longer be able to enter into
those details which had accompanied his account
of preceding ages. The subject became too
vast, and during the seventeenth, eighteenth,
and nineteenth centuries, the number of authors
multiplied to such a degree, that it was impossible for him to do more than select the most important, and he was obliged to divide science
itself into several branches, in order to be more
easily comprehended. The first branch thus
noticed was anatomy, the progress of which he
traced to the middle of the seventeenth century.
He in like manner treated zoology, and the travels which threw light upon it. He then proceeded to botany, mineralogy, and chemistry,
bringing each down to the same period.

The discoveries of Galileo and Descartes were

considered in the eleventh lecture of the second
course, and the influence they and their writings
shed over natural sciences. To this influence
may be attributed the formation of the different
academies of science, the history of which, together with that of the celebrated men who composed them at their commencement, formed a
most interesting lecture. Then, having proved
by cited works and discoveries, that the seventeenth century was the great era of science, and
having finished the history of this period in all
its scientific bearings, M. Cuvier closed his second
course by summing up all that had been said in
an abridged form.

The third course began with the eighteenth
century, which, like its predecessors, passed in
review, though, from its importance and activity,
it, in several instances, required even more division into parts, and various features of it demanded especial notice. To Buffon, for instance,
M. Cuvier devoted two entire lectures, which
at the time were thought to be the most beautiful and eloquent he had ever delivered. This
third course was interrupted from the preceding
Easter till the December following, when he re-
opened it for the purpose of continuing his history from the time of Buffon. He first gave a

clear and eloquent résumé of the philosophy of
Kant, of Fichte, and of Schelling; and one day
in every week was set apart by him, notwithstanding his increased duties as a peer of France,
for the continuation of this immense undertaking. The interval of repose which followed, and
which was absolutely necessary for his health,
was prolonged much beyond his calculation by
the dreadful visitation of the cholera; but on
the 8th of May, 1832, he again resumed the
chair with one of his most impressive and elevated discourses. Never had he spoken with
more fire, nor with more ease to himself: he
" could have continued for two or three hours
longer," he said, " had he not been afraid of
tiring his audience." But they had heard him
for the last time, and this lecture, the memorable
words it contained, and the effect it produced,
seem to me to be so inseparable from his death,
that, for a further description of it, I must refer
the reader to the last portion of this volume,
where the sad details of the closing scene are
related at length. And now having endeavoured,
though I fear but with inadequate success, to
describe M. Cuvier's scientific labours, I cannot
do better than return to that part of his works,
which it is here the principal object to illustrate.

The two examples offered of his familiar style of writing, belong to his private character ; and, in the first, written to Madame Cuvier immediately after starting for one of his journeys, the man, the husband, and the father, are so simply and and unconsciously exposed, that I cannot be too thankful for the permission to make it public. The second was addressed to M. Valenciennes, during the last illness of M. Cuvier's daughter, and both speak too forcibly for the writer to require any further comment.

Pont Sainte Maxence,
Dimanche, 19 Mai, 1811.   Soir.

Ma tendre amie,

Le temps, les chemins, les cheveaux et les postilions se sont trouvés si excellens, que nous sommes arrivés à Pont Sainte Maxence avant six heures, et que j'ai amèrement regretté les deux ou trois bonnes heures que j'aurais pu passer encore avec toi, sans retarder en rien le terme de mon voyage; crois du moins que je les y passe bien en imagination, et que le souvenir de tes caresses, et de ta douce amitié fera le bonheur de toute ma route.   Dis je te prie à Sophie combien j'ai été touché de ses adieux; dis-le aussi à ma bonne Clémentine; pour

Georges, il ne pensait encore qu'au malheur de ne plus avoir des bêtes tous les soirs, mais je te prie de lui en promettre, et même de lui en donner quelques fois de ma part, en bois, en plomb, ou en toute autre matière solide, car il m'a très-bien fait remarquer ce matin, que des bêtes en gravure ne pouvaient pas se tenir debout. Ce pauvre enfant ne se doute pas combien il pourrait rencontrer chaque jour des bêtes qui se tiendraient debout. Ma bonne amie, nous nous portons bien ; nous avons parcouru un pays agréable ; nous sommes dans un auberge supportable ; notre voiture parait vouloir résister : ainsi jusqu'à ce moment tout s'annonce bien. Prie Dieu que cela dure : tu es si bonne qu'il ne peut te refuser. Adieu. Mille tendres baisers.

G. C.

LETTER I.

Ponte Sainte Maxence,
Sunday evening, 19th May, 1811.

My tender Friend,

The weather, the roads, the horses, and the postilions, have proved so excellent, that we have reached Pont Sainte Maxence before six o'clock ; and I have bitterly regretted the two or three good hours that I could still have passed with thee, without in the least retarding the end of my journey. At least, believe that I have passed them in my imagination, and that the remembrance of thy caresses and tender friendship will form the happiness of my whole way. I beg of thee to tell Sophie how sensible I was to her adieus ; say the same to my good Clementine : as to George, he only thought of the unhappiness of not having any more *bêtes* every evening, but I ask of thee to promise him some, and even to give him some occasionally, as from me, in wood, in lead, or any other solid substance ; for he aptly remarked to me this morning, that the *bêtes* in engravings could not stand upright. The poor child does not think how often he may daily meet with *bêtes* who do hold themselves upright. We are quite well, my good friend; we have traversed an agreeable country; and we are in a tolerable inn. Our carriage appears to be quite able to bear the journey; thus, up to this moment, all goes well. Pray to God that this may last; thou art so good that he cannot refuse thee. Adieu. A thousand tender kisses.

Q

LETTER II.*

My dear Friend,

You have done well to go to Leyden, as you will there collect new materials ; besides, at this moment you would only see a spectacle of desolation. My poor daughter is very ill ; and alarm and affliction torment me too much to allow me to devote myself to any regular occupation. Take care of the autumnal fevers. Give my compliments and thanks to M. Temminck. Adieu.

* See fac-simile.

PART III.

I AM now arrived at that part of the Baron
Cuvier's labours which is least known in this
country, and certainly the least understood, on
account of the marked differences which must
always exist between the legislature of two nations so dissimilar in feeling and character as
England and France.　Before I enter upon this
subject, however, I must request my readers to
bear in their memory these three things : —First,
that the improvement of the human mind and
morals was the Baron Cuvier's sole and real
ambition ; secondly, that his leading inclination
was the advancement of science, which he considered the best auxiliary of his views on mankind ;
and thirdly, that the great maxim and rule of
his life was order.　Whatever tended to de-
range these was avoided by him with the most
scrupulous care ; whatever tended to their advancement was most cherished by him.　He
loved his places, because they gave him the
power of executing his great and benevolent

views, and he preferred that mode of government which lent most aid to his enlarged and
philanthropical schemes. At the same time, he
steadily and firmly rejected every thing which
would have disturbed that internal repose of conscience which was absolutely necessary to the
exertion of his own powers.

It is not to be supposed, because M. Cuvier
supported every government under which he
lived, defended its laws, its institutions, and its
existence itself, in his temporary office of Commissaire du Roi*, as Counsellor of the University, and Counsellor of State, that he was blindly
attached to existing forms. On the contrary,
he wished for, he sought amendment and correction; but his knowledge of the history of all
nations, the experience of his youth, taught him,
that the sudden subversion of these forms and
institutions produced anarchy and confusion,
and stagnated every thing like progress; and
what he demanded was, that every attempted
improvement should be the result of deep
thought, calm discussion, and vigorous search
after the necessity of its taking place. He felt
that the passion for innovations of all kinds,

* The office of Commissaire du Roi is, to defend all the
bills brought before either House by the ministry.

which characterised the times in which he lived, produced a constant change of systems, which was calculated rather to destroy than to improve, and, consequently, his actions and counsels were conservative, yet progressive. "He was always the mediator between the time passed and the time to come—between France and other nations; he resisted the antipathy of his countrymen against those whom they chose to call barbarous; and with his whole force always tried to stem the torrent which their vanity and versatility occasionally poured over that which was wise and useful."

It has frequently been remarked, with great bitterness, that M. Cuvier held more places than any man had a right to monopolise. The best answer to this attack is, the manner in which he fulfilled the duties attached to them; a fact easily ascertained now they have passed into other hands, though his career alone can show, how the income of the statesman furnished the savant with the means of carrying on his labours; how the counsellor of his sovereign protected the naturalist; and how "the new Aristotle became his own Alexander."

It would be difficult to decide in which part of his public life Baron Cuvier's talents were most pre-eminent; the affairs of the University alone would have sufficed for most men; for not only were the letters, notes, and remarks which proceeded from his pen in this service innumerable, but, besides these every-day labours, of which the heads of the departments only can form a just calculation, he wrote a mass of Memoirs and Reports, either to enable the directing ministry to comprehend the nature of this institution, or to furnish them with arguments for its defence against its many enemies. Appointed to be one of the members of the Council of the University (1808), he soon attracted the notice of the Grand Master, Fontanes, who named him Commissaire of a discussion about to take place in the Council of State, in the presence of Napoleon, respecting the Imperial University. M. Regnault de St. Jean d'Angely, who spoke against the university, supported his opinion with much warmth, and with all the talent he so eminently possessed. M. Cuvier replied to him, and Napoleon, who had listened to both with the greatest attention, turned towards M. Regnault, and said, " Je crois que vous êtes atteint et convaincu d'avoir tort," &c.* This circumstance, and the reports

* I believe that you stand impeached and convicted of being wrong, &c.

made by M. Cuvier after his return from Italy
and Holland, led the Emperor to appreciate his
legislative talents, and to appoint him Maître
des Requêtes * in the Council of State. His
high opinion went still further; for he ordered
M. Cuvier to select a library for the use of the
King of Rome in his education. The list was
made, and laid before Napoleon at the Thuilleries, when the expedition to Russia put an end
to all these projects.

Raised to the rank of Counsellor of State †
in 1814, M. Cuvier's powers of defence were
constantly called forth in favour of the body
of which he formed a part; and not only did he
shield it from the attacks made upon it, but he
was often obliged to teach the very ministers the
part it played in the government, and the importance to themselves of preserving this institution. The ministerial archives of France

* The office of the Maîtres des Requêtes is, to examine
all questions about to be brought forward in the Council of
State, to report upon them to the Council, and to give their
own opinions concerning the matter.

† This appointment astonished several of those who were
about the Court, and one who was allowed to converse with
Napoleon having asked him why he called a savant to the
Council of State, the Emperor replied, " that he may be able
to rest himself sometimes ;" well knowing, that to a man
like M. Cuvier the best repose was a change of occupation.

contain many of his Memoirs on this subject, by which he demonstrated the necessity of separating the judicial from the administrative part of government, as ordered by the Constituent Assembly; at the same time, that this could not be effected without the creation of a Council of State. The duties of this body are, to prepare laws, to examine ordonnances, and to decide whether the complaints brought against the authorised agents of the government require judicial proceedings. It is composed of enlightened men, who offer a better chance of impartiality than if they themselves were attached to the offices filled by the offending parties. In a very few years after he was admitted to the Council of State, we find M. Cuvier appointed President of the Comité de l'Intérieur *, and from this time his legislative duties were so mingled with those belonging to the University,

* A committee belonging to the Council of State, especially appointed to advise with the Minister of the Interior on all administrative questions, to draw up the ordonnances issued from that body, and to prepare the plans of various laws. This committee examines all the disputes which arise between individuals and the administration, authorises the grants of mines, the construction of bridges and roads, superintends the statutes of different societies, and judges if it be advisable to accept legacies or donations to public establishments, the clergy, &c.

that it becomes difficult, and, in fact, almost impossible, to speak of them separately. Called to these important charges when all required to be revived and reorganised, it is scarcely possible for us to conceive the difficulties that were presented to him: but with what vigour and talent did he put all into action! Public Instruction being attached to the Presidency, he was obliged to draw out the plans for study; to regulate the discipline of the schools; to decide according to the actual necessities of a new order of society; and, nevertheless, only to obey these necessities so long as they did not interfere with those principles of public or domestic order, without which there is no repose, either in a family or a state: in short, to give the rising generation the knowledge and habits most calculated to preserve the great ties of society, and to select those who were most worthy of disseminating such knowledge into every part of the kingdom.	How vast then must have been that capacity which, besides these duties, embraced every branch of science and literature! I dare not dispute that others may have been equally gifted by a beneficent Creator, but I dare affirm, that the one ruling principle of order was the human agency by which M. Cu-

vier brought his heaven-born faculties into full
force.

M. Cuvier greatly occupied himself with
municipal and provincial laws, and those relating
to public instruction; every branch of which was
the object of his exertions. His projects were
often too much modified before they were executed, for the Jesuits, as a matter of course,
were his formidable enemies. Not contented
with issuing ordonnances from the Department
of the Interior, he composed a great many Memoirs to accompany them, which exposed their
motives, and formed so many precious commentaries, as they explained with the greatest perspicuity the reason of every article. He thought
it as useful to spread every where the reason of
the laws as to disseminate the laws themselves;
thinking that the latter are often attacked and
mistaken by the public for want of a proper comprehension of the motives which caused them to
be framed.

Under the ministry formed on the 26th of
September, 1815, and composed of the Duc de
Richelieu, Marbois, Corvetto, Fittre, Vaublanc,
Dabouchage, and de Cazes, M. Cuvier was enabled
to render an essential service to France, which
I cannot do better than describe in a translation

of his own notes. " I had then an opportunity of rendering great services to this country, which have never been publicly declared, but which I should be sorry should not one day be known to have emanated from me. R———— supported me in all the ameliorations we brought about in the Council concerning the criminal laws, which were prepared in the spirit of the times, but the modifications which rendered those of the Pre-votal Courts * almost inoffensive are due to me. In the first place, judicial power was given to them, not only over revolts, and attempts openly committed on the public peace, but over conspiracies and attempts plotted in secret; and not only over crimes which might take place after the law was promulgated, but over all which had taken place at any period whatever.   It is very evident that in a country like ours, where there are so many men of all classes ever ready to follow the torrent of the day, these two powers would have transformed the Prevotal Courts into so many revolutionary tribunals.   Nevertheless, we did not obtain any thing from the united Committees

---

* The Prevotal Courts were created by the Bourbons, in order to judge all public disturbances, and from whose decisions there was no appeal.   They in some sort assimilated to our special commission.

of the Interior, and the law was prepared; but after a meeting of the Council of State, presided by the Duc de Richelieu, I demanded a discussion of these questions in his presence before a new assemblage of the Committees. I believe that I never spoke with so much fire; and, notwithstanding the violence of —— and ——, thanks to the upright and honest mind of the Duc de Richelieu, I succeeded in getting the articles concerning secret plots entirely erased. There yet remained the visitation of former offences to be overcome. M. de —— opposed it in the Committee of the Chamber of Deputies, where it was defended by two counsellors of state; I was invited to join them, as I should naturally have been obliged to do in my office of Commissaire du Roi, but I refused, and the law did not pass. The Prevotal Courts have already caused evil enough by the manner of their establishment, but I venture to affirm, that their mischievous effects would have been incalculable if the plan had not been changed on these two points. I am the sole cause with respect to that of the secret plots, but with regard to the punishment of past offences, M. de —— contributed with me to its being abandoned."

Always guided by the feeling of the good he

could effect, and the evil he might avert, under every change of ministry M. Cuvier was to be found, not only defending the institutions which were in danger of being overthrown, but in the Chambers and in the Council, generally successful in preventing those alterations which would have reduced the objects of his unremitting cares to a state of feebleness. Under the ministry of M. ——, a proposal was made to introduce the Jesuits into the University, or, in fact, to deliver it into their hands, and M. Cuvier's firm and spirited resistance alone prevented this measure, which, in all probability, would have caused its destruction. His refusal to form a part of the commission for the censorship of the press, at a moment when, from the despotic nature of the government, this refusal might have been followed by the most grievous consequences to himself, yet more forcibly proves that he was not the man to preserve his places at the price of his reputation. As this occurrence has been much misrepresented, I shall relate all the circumstances which attended it. In M. Cuvier's capacity of Counsellor of State, he had been one of the first most vigorously to oppose the censorship, and fearlessly maintained his opinion, both in full Council and in the Chamber of Deputies ;

using all the energy and reasoning he could
command, and leaving nothing undone to put a
stop to the measure. Thus far he had only to
act in strict accordance with the rank he held in
the state, but the interference of another body
placed him in a more painful situation. The
Académie Française, of which he was so distinguished a member, determined to interfere in
this question; and it became a matter of consideration and dispute, whether a purely learned
assembly had any right to join itself to party,
and intermeddle with affairs of state. M. Cuvier
was of opinion, that it entirely lost sight of its
proper character by so doing; that it would
thus endanger the harmony of the members
among themselves; that it destroyed its opportunities of usefulness by not retaining its inde-
pendence of politics, and completely overstepped
the legal boundary, by presenting a petition from
a body, which privilege in France is only accorded
to individuals. These motives alone (for he dared
not listen to the detestation he felt for the cause
of this step on the part of the Academy) induced him to employ all his eloquence to prevent
the petition from being presented to the King.
He unhesitatingly exposed the inconveniences, the
hateful bearings of such a law; but he persisted

in it, that the Academy had no right to mingle itself with political questions ; and that, if it once suffered itself to assume such a privilege, it would at length dwindle to a mere instrument of party. On this occasion, however, his eloquence and reasoning proved of no avail; the King was petitioned by the Academy, but Charles X. would not even receive the deputation. The rejected dignitaries found favour with the multitude, and, of course, M. Cuvier, and those of his opinion, were accused of pusillanimously preserving their places at the expense of good feeling. The project, however, owing to the resistance of the Chamber of Peers, which then possessed more weight than at present, was for a while abandoned. In the same year it was renewed, and, without even asking his consent, before he was in the least aware that the measure had been decided on, the ministry appointed M. Cuvier one of the censors of the press. On Sunday the 14th of June, 1827, at midnight, arrived an official despatch from the government, written by M. Peyronnet, to announce to him that his appointment to this office would appear the next morning, at nine o'clock, in the Moniteur. To refuse the intended honour; to foresee the probable consequences of such a refusal ; to yield to these consequences without hesitation; in short, to prefer conscience to interest, was the work of an instant; and in ten minutes, a firm, dignified, but moderate refusal, was sent to the Chancery. The ordonnance was at that moment printed, and M. Cuvier's name appeared in the list of the morning, because it had been physically impossible to get it erased; but private means were taken to publish his refusal in every quarter, till all France was in entire possession of the fact. Most of the papers, under fear of the censorship, had been unable to repair the error; and, in fact, when the Journal des Debats ventured to insert M. Cuvier's rejection of the office, the lines which contained it were scratched out by the censors.* This conduct, with the fickle public, regained M. Cuvier's lost popularity, but produced great coolness towards him on the part of the King. I ought not, however, to omit mentioning, to the credit of Charles X., that this coolness ceased after M. Cuvier's dreadful calamity. The first time he appeared at Court after the loss of his daughter, his Majesty addressed him with considerable feeling and

* Others refused the office at the same time, but I have only to speak of M. Cuvier.

kindness, asked him several questions relative to the event, and expressed himself as deeply concerned.

For the last thirteen years of his life did M. Cuvier preside over the Comité de l'Intérieur, and the number of affairs which passed through his hands in this office alone is almost frightful to the imagination: I ought not, perhaps, to say passed through, but that they were examined, deeply considered, and forwarded by him. I should speak much within the limits of the truth, if I were to state them at ten thousand every year. The art of properly distributing the work among his colleagues; his talent in directing discussion; his unfailing and prodigious memory, supplying antecedent decisions at the desired moment; his profound knowledge of the principles which ought to regulate each affair, the best method of applying these principles at the best opportunity;—these qualities all rendered his presidency the most remarkable of the present age, and have indelibly impressed it on the recollection of all who had the advantage of labouring with him. To see him at one of these meetings was, perhaps, to see him in his greatest perfection as a legislator. Rarely eager to give his advice, he even appeared to be thinking

R

of subjects wholly irrelevant to the matter in discussion; but he was often, at that very moment,
writing the judgment or regulation which must
necessarily follow the deliberation. His turn to
speak only came when all others had stated their
reasons, when useless words were expended.
Then a new light burst upon the whole; facts
assumed their proper position, confused and mingled ideas were arranged in order, the inevitable
consequences appeared, and when he ceased to
speak the discussion was terminated."*

But these were not all the legislative labours
of M. Cuvier. Always holding the office of
Chancellor to the University, he had twice been
forced, in the temporary vacation of the Grand
Mastership, to take upon himself the highest
dignity, and, during these two periods, fewer
complaints were made against this institution
than at any other.† A most gratifying proof of

---

* These are nearly the words of one of M. Cuvier's brother
legislators, the Baron Pasquier, to whose eloquent éloge, delivered in the Chamber of Peers, of which he is president, I
am deeply indebted. My sole object is to do justice to
M. Cuvier's talents and character; and to accomplish this, I
may be excused for employing better language than my own,
especially when the writers speak from personal knowledge.

† It should be understood, that, in twice accepting the
functions of Grand Master for the time being, M. Cuvier
never received the salary attached to this high dignity,

the respect felt for him took place at the moment that a change was made in this appointment. It was the duty of the newly named
Grand Master to distribute some prizes awarded
by the University; but he was very far from being
popular, and as the public mind was at that moment considerably agitated by political events,
it was generally understood that the students
intended to raise a violent commotion. Whether
he was in reality unprepared with his discourse,
or whether he feared the consequences of ap-

---

though it increased his household expenses, and though it
was richly endowed, even under the restored government.
The following are the dates of his holding this office :—
M. Royer Collard vacated the presidency of the committee of
public instruction on the 13th of September, 1819.   M. Cuvier replaced him on the same day; and a letter from the
minister of the interior, dated the 17th of the same month,
notified the desire of the King, that the committee should
continue its labours under the presidency of the counsellor
holding the place of chancellor.   M. Cuvier was that chancellor, and continued in the rank of Grand Master until the
21st of December, 1820, at which period M. Corbière was appointed to it.   M. Corbière resigned on the 31st of July, 1821,
and M. Cuvier was again chosen to fill the vacancy until the
1st of June, 1822, when M. Frayssinous was named Grand
Master.   The day on which M. Frayssinous was called to
the ministry, M. Cuvier was appointed Grand Master for
Protestant Affairs, which dignity only ceased with his existence; and, let it be remembered, was equally filled by him
without pecuniary remuneration.

pearing on this occasion, the recently chosen
dignitary requested M. Cuvier to officiate for
him.   Tottering, as the University was at that
time, under the enmity of many powerful men,
one act of violence, one instance of excitation
and imprudence on the part of its youthful followers, might have caused its downfall; but M.
Cuvier met the difficulty with his wonted energy
and judgment.   Half an hour, taken from the
duties of the preceding evening, was devoted to
the composition of a discourse, which only required some minutes to deliver.   The day arrived, and the students appeared, manifesting
every hostile disposition.   The sight of M. Cuvier first checked their excited feelings: they
remained in respectful silence; the reason and
gentleness of his expressions restored complete
tranquillity; the distribution took place, and, as
the benevolent and revered master laid the
crowns upon the heads of his disciples, he addressed them as a father would his children;
nothing but the murmurs of gratitude and applause were heard, instead of the angry and tur-
bulent conduct threatened, and the University
was saved.

Even had Charles X. made the Grand Mastership of the University a permanent office,

which intention he more than once expressed,
M. Cuvier could not have held it, owing to the
insurmountable obstacle presented by his religion; therefore was he made perpetual Grand
Master of the Protestant Faculties. This honour was not even opposed by the Catholic
bishops, who were thoroughly sensible of M.
Cuvier's profound knowledge of ecclesiastical
affairs, of his tolerating spirit, which never
prompted him to one harsh measure, and he
assumed this important charge to the great satisfaction of all France; if, indeed, we except
the fanatics of his own creed, who were, perhaps, as much opposed to his enlightened views
as the Jesuits, and caused even greater obstacles
to the ameliorations he endeavoured to establish.
He instantly commenced a completely new system of order and encouragement, which, it is
true, did not always succeed according to his
hopes; so difficult is it to wean the unenlight-
ened from the prejudice of party, and from long-
established ideas. He, however, perseveringly continued his endeavours, undaunted even by the
failure of many schemes, provided he could be
successful but in one instance. He believed that
instruction would lead to civilisation, and civilisation to morality; and, therefore, that primary

instruction should give to the people every means of fully exercising their industry without disgusting them with their condition; that secondary instruction should expand the mind without rendering it false or presumptuous; and that special instruction* should give to France, magistrates, generals, physicians, clergy, and professors, all distinguished for their enlightened views; in fact, that succession of elevated characters which make the real and imperishable glory of the country in which they act their part. But here it may be interesting again to introduce M. Cuvier's own words, as expressing his sentiments, and which have been supplied to me by M. Laurillard:—" Give schools before political rights; make citizens comprehend the duties that the state of society imposes on them; teach them what are political rights before you offer them for their enjoyment. Then all ameliorations will be made without

---

* These three terms of primary, secondary, and special, to which I believe we have nothing analogous in England, designate, first, the instruction given to the poorer classes both in town and country, and which, in France, is confined to reading, writing, and the first four rules of arithmetic; secondly, a more extended education, fit for general purposes; and, thirdly, a still more elevated course of study, which fits the pupils for any particular career to which they may direct their views.

causing a shock; then each new idea, thrown upon good ground, will have time to germinate, to grow, and to ripen, without convulsing the social body. Imitate nature, who, in the developement of beings, acts by gradation, and gives time to every member of her most powerful elements. The infant remains nine months in the body of its mother; man's physical perfection only takes place at twenty or thirty, and his moral completion from thirty to forty. Institutions must have ages to produce all their fruits; witness Christianity, the effects of which are not yet accomplished, notwithstanding a thousand years of existence."

With such objects always in view, M. Cuvier attempted and executed several improvements, of which I shall now speak. The buildings of the ancient Collège du Plessis, in which the faculties were placed, being in a state of general dilapidation, he obtained from the government, a grant of the Sorbonne for their use; and as it was highly important that the lectures should not be interrupted during the removal, he exerted all his activity, incessantly visited the architect appointed to direct the works, and reiterated his own inspections, till the object was accomplished. The Faculty of Sciences owe

the funds they possess for a cabinet of natural
history, and for the purchase of various instruments, entirely to M. Cuvier's efforts. The appointment of medical officers, who understood
natural history, to the government vessels, was
solely due to his suggestions, as well as that of
attaching collecting travellers to the museum of
the Jardin des Plantes. The treasures brought
home by the Uranie, the Coquille, and other ships,
are proof of the excellence of the first plan; for
the officers were delighted to employ their leisure
in drawing, describing, and preserving the objects
they met with in the course of their expedition.
The rapid increase of the museum at the Jardin,
during the life of M. Cuvier, speaks too plainly
for the latter to need further comment. The
mode of appointing professors is a complicated
question in France : some are partisans for election by vote, some support nomination by established high authority, and others, succession.
Each of these methods is attended with inconvenience; and voting, which theoretically may
appear to be the best, has not realised the
hopes of those who caused it to be adopted. It
gives an opportunity for all to enter the lists;
and men of consummate skill and experience
do not like to find themselves placed in contact

with those just issued from the schools; who, with all the fire and confidence of youth, frequently obtain their wishes by their brilliancy, while those of much more real merit are left far behind. The other methods are particularly open to private feelings, or a liability to place men of inferior merit in the professor's chair. To obviate these inconveniences and abuses, M. Cuvier created that method which, in France, is called aggregation. A defeat is of comparatively little consequence to young students; and therefore, according to this institution, directly they quit the schools, they undergo an election in order to become agregés : these agregés are assistants to sick or aged professors, during their attendance on whom, time and opportunity are given for the developement of their talents, and to make themselves known. At the death of the professor, the faculty to which he belonged presents three candidate agregés to the minister, whose choice determines the appointment.

Long convinced that those destined to different administrative functions should follow a course of study especially adapted to these duties, in the same manner as they do for the learned professions, M. Cuvier proposed to the Simeon ministry to create a new faculty, or par-

ticular school of administration, on the models
of those which have long existed in Germany,
and to which faculty he desired that his name
should be attached. The project was prepared,
and nearly put in execution, when the above
minister retired, and from the succeeding party
nothing further could be obtained than a chair
for administrative law, and even that was very
shortly suppressed.

All the minor schools of France were likewise
the objects of M. Cuvier's earnest solicitude;
and, notwithstanding the frustration of many of
his plans, from an obstinate attachment to old
methods, he succeeded, by reiterated appeals to
the government, in establishing among them
professors of history, living languages, and natural history. In order to further primary
instruction, he caused the institution of provincial committees for superintending the schools of
their own departments, thinking that emulation
would thus be excited among those called to the
office, consequently their zeal redoubled, and
their endeavours carried to a greater extent. In
some provinces this plan was attended with the
greatest success, but in others party spirit and
consequent dissension paralysed even the most
active. As a further proof of the ever-watchful

cares of M. Cuvier, and the extent to which he carried his anxious endeavours, I now quote a discourse delivered after his death. In this we have not only the words of the superintendent of an important establishment, but his own in evidence of the extreme interest he felt in this primary branch of education. This discourse emanated from M. Reynal, Rector of the Academy of Bourges, and was delivered at the distribution of prizes in the protestant school of Arnières.

" My dear children,— The faithful followers of our church have spared sufficient from their own wants to build a school for you, and to afford you all that is desirable for your instruction. The academy (of Bourges) has associated itself with them in this work of devotion and self-sacrifice; it has already done much, and will do still more, by appealing in your behalf to the benevolent protection of the university. You see, my dear children, that you have friends and benefactors every where. But, alas! he is no more— he who held the first rank among them! A premature death has snatched him from science, from literature, from your brethren, from us, from all mankind. The whole of the learned world deplores his loss. You are too young,

my children, to have heard him spoken of in
your village, but the great man who tried to do
you so much good, who unceasingly thought of
you, was called George Cuvier.   Recollect this
name, and mention it every day in the prayers
that you address to heaven.   He has often
written to me, ' Do not, Monsieur le Recteur,
lose sight of our school at Arnières les Bourges.
I recommend the scholars of it to you as my
brothers, as my best friends.   Instil into them
submission to their parents, respect for the property of others, candour and justice.   These
are the virtues and duties of all religions.   Let
benevolence and affection reign between them
and the children who inhabit the same village,
and who, like them, live by their labours in the
fields.   God loves and protects them all with
equal goodness; with the same hand he blesses
the sweat of their brow, and their harvests; let
them, therefore, behave towards each other like
children of the same father.'

" My conscience tells me, that towards you,
my children, and all the young people confided
to my care, I have fulfilled my duties, and most
especially the views of your protector, the
great man whom we lament.   This place is very
humble, my voice is very feeble for the praise of

such a life. The eulogium of George Cuvier,
of the same religion as yours, will often appear
in your books, and will be pronounced by our
learned bodies and our most celebrated orators.
However, a word sincerely uttered within these
walls to the memory of one so dear to us, of the
learned and virtuous man who deigned to
honour me with his goodness, has not appeared
to me to be out of place, or without interest; and
it is as much for your sake as for my own that
it becomes me here to speak of George Cuvier,
and to lead you in him to preserve the recollection of one of your most ardent benefactors. Let
this short eulogium serve you as a lesson, and
teach you to be always grateful to those who
wish us well, and especially to those who do us
good."

During M. Cuvier's direction of the Protestant Faculties he became one of the Vice-Pre-
sidents of the Bible Society, and caused the
creation of fifty new cures, which had very long
been wanting. The protestant churches required fresh regulation and discipline, and for
this purpose he collected the opinions of the
different pastors of these churches, placing in this
matter, as well as in all others, great confidence
in the counsels of experience; and had, in con-

sequence, drawn up the plan of a new law,
which was to have been laid before that session
in which he did not live to take his seat. The
feeling with which the ministers of his own religion generally viewed him will be proved by
the following extracts from the discourse delivered at his funeral by M. Boissard, minister
of the protestant church in the Rue des Billettes.
" Let us not forget those long abandoned cha-
pels re-opened to our youth in the royal colleges; let us not forget the abundant distribution
of religious and moral books under his superintendence. Now that his voice is extinct, let
us fervently ask of our God, let us ask in the
name of our dearest moral interests, in the name
of our eternal welfare, to raise up other voices,
which may speak with the same eloquence, the
same wisdom, and the same authority. We have
lost him who, with inviolable attachment, honoured the creed of our forefathers; whose
great name, whose immortal labours, shed so
much lustre over our churches; who burdened
himself with our ecclesiastical rights in perfect
disinterestedness of spirit, and with the purest
and most extensive benevolence. What do we
not owe to that penetrating glance which revealed to him all that was wanting in our insti-

tutions, and under which privations we had so long groaned! How many ameliorations took place in a few years; with what wisdom and charity he examined our requests; and what a new order of things would have arisen at his bidding, had the Almighty suffered him to continue among us!"

M. Cuvier's elevation to the Chamber of Peers was but a just tribute to his long and important services; and he took his place among his new colleagues with that calm dignity which was not likely to be ruffled by any accession of rank. It was wholly unsolicited, and, at the moment, produced any thing but joy in himself or his family; for it appeared likely to draw him still more into public life, at a period when all around was stormy and uncertain. It is well known how the Chamber of Peers felt it their duty, for the preservation almost of their existence as a body, to vote in the agitating question of inheritance *; and, of course, M. Cuvier acted according to his maxim of preferring the lesser evil, when evil was unavoidable; but, when unshackled by

* In case of continued opposition, the Chamber of Deputies threatened to form themselves into a Constituent Assembly.

such imperious circumstances, he defended the
University, and various questions of finance, in a
manner which showed how little he courted popularity. The only work of his hands which
remains in the archives of this Chamber relates
to corn laws, and was written in a very few
hours. But these few hours reflection, on so
arduous and delicate a subject, sufficed him for
the production of an exact and extensive statement of the facts which rule these laws in
France, of the facts which ought to rule them,
of the legislation applied to them during a certain number of years, and, lastly, the considerations which operated in favour of the measures
proposed in the Chamber, and which were finally
adopted.

A remarkable proof of the comprehensiveness
of his legislative talents occurred in this Chamber during his short career there. A question,
purely military, was discussed, and so many arguments took place that the affair became confused, and resisted all the efforts made to come
to an explanation. M. Cuvier rose, immediately
set the whole in a clear, strong point of view, so
as to enable the desired arrangements to take
place ; and this, not in consequence of any de-

duction made from the reasoning of the moment,
but from a thorough knowledge of the matter in
all its bearings.

The loi de cumul (law against accumulation)*
would have deprived him, had it been enforced,
of one third of his income; but it was contemplated by him with the utmost calmness, and,
even at the moment when the enforcement appeared to be inevitable, he prepared himself to
continue in the performance of all his high functions in the state, without the slightest pecuniary
profit.   His family cheerfully adopted his views,
from the feeling that these duties formed a variety of employment, which was a relaxation to
such a mind, and consequently beneficial to his
health.   The loi de réduction† (law of reduc-
tion), which attacked all places, did take away
a considerable part of his revenue; but this excited no other emotion than a regret that it

* This law was to prevent any one man from benefiting
by the salaries attached to a number of places, at one time;
and, in fact, was intended to prevent the holding of an accumulation of employments by any one individual.

† The loi de réduction was to lower the salaries of all
those who held public functions; and, as M. Cuvier was
charged with more than any body else, it, of course, made a
great difference in his annual receipts.

must curtail the exercise of that hospitality for which he had so long been remarkable.

A trifling circumstance happened during the latter part of M. Cuvier's legislative duties, which, as it was erroneously supposed to have been connected with his death, may as well be properly explained here. While defending that incessant object of his anxious cares, the University, before the Chamber of Deputies, in his office of Commissaire du Roi, his voice was much interrupted by a violent cough, on which several of the members came to him, to beg that he would go into the Salle des Conférences, and drink some water: one of the deputies put M. Cuvier's arm within his, and led him so fast, that his foot slipped down a step, and he almost fell to the ground. The hand, however, that had caused the mischief, supported him; but he was immediately surrounded by most of those in the Chamber, who, evincing the deepest interest, obliged him to seek some refreshment in the adjoining room. The strongest proof that no malady had caused this appearance of accident, is, that, ten minutes after, he mounted the tribune for the second time, and, with his usual force and clearness, for more than an hour, once again defeated the enemies of the University. The malady, which in reality laid his mortal frame in the dust gave no warning; and, from the moment it appeared, left no doubt, in his own mind at least, of its destructive result.

PART IV.

In the first part of this volume, I have, by a narrative of the principal events of the Baron Cuvier's life, endeavoured to show his progress towards fame and honour, and to expose the circumstances which tended to the perfection of one of Nature's noblest productions.   In the second, I have, notwithstanding the difficulties of such a task, and a strong feeling of my own weakness, tried to give an outline of M. Cuvier's principal works, of his most important discoveries, and the immense advantages that science has derived from each.   In the third, I have studied to lay before the reader all the good he effected, and all the evil he prevented during his administrative career ; and, by so doing, I have set forth all his titles to the grateful admiration of mankind.   I am now about to attempt a more particular description of the character, the private virtues, and domestic habits of the great man, whom I have so often admired in the midst of his family, surrounded by friends, and performing the honours of his house to a

numerous circle, composed of men of all countries
and professions. But before I enter into these
details, I feel called on to refute, by a reference
to known facts, those accusations which have
but too often been brought against him. Men
rarely pardon superiority, even when (as in M.
Cuvier) it is exempt from all kinds of vanity;
still more rarely do they pardon those acquisitions of rank and fortune which necessarily result from this superiority; and the great number
of places held by M. Cuvier, caused him to be
accused of an ambition for power, by those who
reckoned his employments, without reckoning
his merits, or without recognising how useful his
talents were to France.

In order to set aside this unjust charge, it
will be sufficient, here, to recall some of the
occurrences scattered through these pages; and
which lead me to affirm, that, so far from having
sought or solicited places, he nobly rejected
several which were offered to him. Twice, at
different periods, did he refuse the directorship
for life of the Museum of Natural History, and,
at another, to enter the ministry,— an advancement which at that time no one thought of re-
pulsing; and the greater number of the favours
conferred, reached him during his absence, and

were wholly unexpected. It was during his
journey to Marseilles that the Institute appointed him perpetual secretary; it was while in
Holland that he received from Napoleon, an
endowment, with the title of Chevalier; he was
at Rome when the Moniteur informed him of
his having been named Maître des Requêtes;
in England, when he was elected to the Académie Francaise; it was in the midst of studious
retirement, when he had, as it were, shut out
the world, that the rank of Peer of France
crowned his administrative career; and, lastly,
on the day of his death, his nomination to the
Presidency of the entire Council of State was
presented for the signature of his sovereign.*
It may, therefore, be safely said, that honours
sought him; and now, that his decease has left
so great a void in every institution, in every
learned and administrative body which could
boast of his name on their lists, and most of
which were so powerfully served by his labours,
I trust that his actions, and the noble disinterestedness of his character will be acknowledged,

---

* I do not reckon among these honours the election to
almost every learned body in the two hemispheres; for, of
course, all were anxious to pay so just a tribute to M. Cuvier's
pre-eminence.

and that the breath of envy will no longer dare to mingle with the testimonials of admiration which are to be heard on all sides.

There is yet another sort of reproach, which the inventors of systems overthrown by M. Cuvier have dared to bring against him. These, wounded by self-love, or contradicted in some cherished fancy, have not feared to attribute to pride, or even to a feeling of jealousy, very far from his noble heart, the reserve with which he admitted certain explanations of the phenomena of nature, and the resistance he offered to limited or defective theories, the errors and insufficiency of which, his penetration and profound knowledge instantly led him to discover. This resistance, however, was one of the most beautiful parts of his character, for it proved his love of truth, and the ardour with which he knew how to defend it, even at the expense of his own tranquillity; and he fearlessly exposed himself to personal enmity, in order to turn students away from such views, the inevitable result of which was, to stop the progress of science, by giving a false direction to the minds of those engaged in her cause. Speaking of theories in general, he said, a little before his death, " I have sought, I have set up some myself, but I have not made

them known, because I have ascertained that
they were false, as are all those which have
been published up to this day. I affirm still
more; for I say, that, in the present state of science, it is impossible to discover any, and that is
why I continue to observe, and why I openly
proclaim my observations. This alone can lead
an author to the discovery of that fact on which
he can build a true and general theory.... This
fact," added he, " is perhaps of little importance
in itself; but, with regard to theory, it will become the principal fact, the keystone to the arch.
Therefore it must be sought, science must march;
but we must take care that she does not march
in a retrograde direction, as she has sometimes
done, and as some naturalists lead her at present.
We ought to labour, not with the object of sup-
porting a theory,—because, then, the mind being
pre-occupied, will only perceive that which favours its own views, — but with the object of discovering the truth; because the truth will be
deduced from true theories, and true philosophical principles; the truth being, in itself alone,
the whole of philosophy." *

It seems that both the French and the Ger-
mans claim M. Cuvier as their countryman; and

* M. Laurillard.

it would be difficult to decide whether the place
which gave him birth, or that which was the
scene of his labours, has the best title to call him
her own.* His family, as we see in the previous pages of these memoirs, was originally
Swiss; and, being driven to Montbéliard in consequence of professing the reformed faith, settled there as a remote province of Germany,
and where some of the members of it held important charges. His uncle was a minister of
the Lutheran religion, and his father an officer
in a Swiss regiment then in the service of France.
I am led to dwell on these two circumstances,
from errors committed by several writers, who
have stated M. Cuvier to have entered the
church; and also a mistake made by M. Decandolle, a very old and esteemed friend of M.
Cuvier, and the learned botanist of Geneva.
This gentleman asserts, in his funeral éloge of
M. Cuvier, that the latter entered the army,
which assertion is wholly without foundation;

---

* The year in which M. Cuvier was born was a remarkable one, for in it Napoleon Buonaparte came into the world,
who made as great a revolution in the political face of Europe,
as M. Cuvier did in that of science, though not equally lasting.
The Duke of Wellington, Mr. Canning, M. de Châteaubriand,
Sir Walter Scott, Sir James Mackintosh, alike drew their first
breath in this year.

and it is very probable that both errors have arisen from some confusion between the father, the uncle, and the son.

There is yet another erroneous report, which I am desirous of correcting; and that is, the late developement of M. Cuvier's talents for natural history. So far is the fact contrary to this, that, even while at the preparatory school of Montbeliard, his greatest happiness was to read Buffon, to copy the plates, and to colour them according to their descriptions. When arrived at Stuttgardt, his studies took a higher flight; and he chose that faculty which allowed him to pursue his favourite occupation. As age increased, his boyish pleasure became, as it were, a passion, and he incessantly pored over all the books he could find on this subject. He dissected the only things within his reach, such as insects and plants; he made an excellent collection of the latter, and discovered several species, in the neighbourhood of Stuttgardt, which were not previously known to exist there. He kept a number of living insects in his room, constantly feeding them, and watching their habits. It was there that he made many of the drawings spoken of in Part II., and which form several thick volumes. I have two of these, which show that

the hand of the master guided him even at
this early age. Knowing the great interest he
felt in such productions, in one of my visits to
Paris, I took a collection of original drawings
for his inspection. Every evening during my
stay there, he asked for my book, and one morning entered the breakfast-room with a huge
quarto in his hands, and, putting it before me,
said, " Permit me to enter the company of
your friends : choose any two of these pages, and
I will cut them out for you. I amused myself
with drawing these figures when I was a student
at Stuttgardt ; and if I were to draw them now,
I could not make them with greater accuracy."
This same facility for designing continued
throughout life ; and it is scarcely possible to do
justice by words to his anatomical drawings, in
which he had a manner peculiar to himself of
expressing the cellular tissue. His delineations
of quadrupeds were equally extraordinary ; and,
when lecturing, he would turn to the black board
behind him, with the chalk in his hand, and,
speaking all the time, he would rapidly sketch
the subject of his discourse, sometimes beginning even at the tail, proportioning every part
with admirable precision, and preserving the
character to such a degree, that even the species

could be immediately pronounced. The taste
for drawings of natural history extended to all
branches of the art, and it was his delight to
visit every collection or exhibition of the kind.
During his last visit to England he went to Hamp-
ton Court, and it was with difficulty he could tear
himself away from the cartoons of Raffaelle, in
order to keep a dinner appointment. The admiration he felt for this most wonderful of all
painters amounted to a species of worship; and
no one, whether an artist or not, ever comprehended or delighted in the beauties of Raffaelle
more than did M. Cuvier. His long stay in
Italy had refined and confirmed his judgment;,
and when he was accused of want of proper
curiosity for not extending his route as far as
Naples, during either of his journeys to Rome,
he deemed it sufficient to reply, "At Naples
I should not have found the Vatican!" He was
very sensible to the merits of our great Lawrence, to whom he was personally attached, and
who had constantly sent him the engravings
from his works; and also to the conception and
genius of our Martin, whose engravings had
always excited his attention in Paris, and whose
gallery he visited when last in London. Woe,
however, to the artist who committed a fault in

anatomy or perspective; his quick eye immediately fastened on it, even in the midst of the praises excited by colouring or expression. To view the exhibitions of the works of our celebrated portrait painter, which took place after his death, was one of the objects of M. Cuvier's second journey to this country; and he frequently passed hours in the British Gallery, where they had at that time been collected. He had personally known many of those represented by this life-giving painter; he felt, as he contemplated them, as if he were again in their presence, and related a thousand anecdotes, which he was delighted to recall.

There was yet another talent of M. Cuvier's, which manifested itself in his earliest youth, and which, though trifling in extent, was a further proof of his facility for retaining a recollection of form. It was the power of cutting out, in pasteboard or paper, whatever object had excited his attention; and a remarkable proof, not only of his dexterity, but of his quick perception, occurred when he was about six years of age. A mountebank passed through the village, who played various slight-of-hand tricks, and was called in by M. Cuvier s uncle to amuse the children assembled at his house. A " fontaine

de Héron *," which ran and stopped at his bidding, a poniard which he apparently plunged
into his arm and drew out again, dripping
with blood, amused and astonished the spec-
tators of all ages who happened to be present:
but George Cuvier examined every thing with
deep attention, and evinced little or no surprise:
for he explained the manner in which the fountain played, and the mechanism of the poniard,
accompanying his explanations by cutting the
whole out in paper.

But I must beg my readers once more to go
back to Stuttgardt, where M. Cuvier obtained
honours which were conferred only on the chosen
few, and those few much older than himself.
His first examination at that university had something remarkable in it, considering that he was
then but fourteen. The committee deputed to as-
sign him his place, reported of him as follows :—

" The young Cuvier has shown, 1st, just notions of Christianity, well adapted to his years.
2dly, A good knowledge of general history and
geography. 3dly, Solid notions of logic, arithmetic, and geometry. 4thly, Considerable skill

---

* So called by the French, because it was invented by
Hero, of Alexandria, who lived 120 years before Christ.
Its English name, I believe, is " a fountain of circulation."

in making Latin themes and verses, and in reading the New Testament in Greek." At the
moment of entering the academy, he was ignorant of German ; but, as we have already seen,
in less than a year, he secured the prize for
that tongue. He always retained the faculty
of speaking this language, to which he added
Italian, in both of which he conversed fluently.
He read several others, and, among them, English; his inability to speak which, I have often
heard him regret. He was accused of knowing
more of it than he chose to own ; but there could
be no motive for concealing what would have
afforded him pleasure to make use of; besides
which, he has often tried to put little sentences
together in jest, and ask if they were right. If
a reply was given in the affirmative, he would
threaten to begin in earnest one day, and probably would have performed his intention, had not
his daughters always acted as able interpreters
in this respect. His knowledge of the dead languages was not only a source of exquisite enjoyment to him, but was absolutely necessary to his
profound researches. He seldom alluded to
Greek or Latin authors in conversation, but
there was a classical precision and elegance of
expression, even in his ordinary discourse, which
can scarcely be acquired by other means than the

study of such writers. The minor accomplish-
ments which he added to these mental stores are
almost surprising, because each must have taken
time to acquire. Among them was a thorough
knowledge of heraldry, which, it is well known,
contains a large portion of detail.

There cannot be a stronger proof of the pre-
cocious perfection of M. Cuvier's universal talents than the occurrences of that part of his
life which was spent in Normandy. One or two
of these (in addition to those already mentioned)
I have extracted from the eloquent éloge delivered by Dr. Pariset at the late meeting of
the Institute.* "A citizen of Caen, who was
a great amateur of natural history, possessed
a magnificent collection of the fishes of the
Mediterranean: the instant M. Cuvier heard
of it, he flew to inspect the treasures, and,
after several visits, he, by means of his pencil,
that precious instrument of observation and memory, became in his turn the possessor of the
collection; for, in natural history, the faithful
representation of an object is the object itself.
Nearly six years passed in this manner, terribly,

* Dr. Pariset is one of the physicians to the Hospital of
La Salpêtrière, and, as Member of the Academy of Medicine,
composed and read the above éloge, which was heard with
the most reverential attention, and followed by enthusiastic
applause.

indeed, to France and Europe, but calmly and profitably to M. Cuvier. Nevertheless, the Revolution insinuated its jealousies and suspicions even as far as his abode; and, the impulse having been given from the capital, one of those societies, or unions, was about to be formed at Fécamp, which armed the people against themselves, and were attended with the most injurious consequences. M. Cuvier saw the danger, and represented to the owner of Fiquainville, and the neighbouring landholders, that it was to their interest to constitute the society themselves. This wise counsel was adopted; the society was formed; M. Cuvier was appointed secretary; and, instead of discussing sanguinary politics at its meetings, it devoted itself solely ' to agriculture.' I have already related how M. Tessier happened to have taken refuge in the neighbourhood, and how he was detected and accosted by M. Cuvier; I have now to add, from M. Pariset's éloge, that, after this greeting, they became the greatest friends; ' and that the perfect confidence which existed between them, in a measure, rendered them necessary to each other.' M. Tessier daily discovered in his young friend new talents and perfections, and was astonished at the sight of his numerous pro-

ductions. On the 11th of February he wrote
as follows to M. de Jussieu:—' At the sight of
this young man I felt the same delight as the
philosopher who, when cast upon an unknown
shore, there saw tracings of geometrical figures.
M. Cuvier is a violet which has hidden itself
under the grass; he has great acquirements, he
makes plates for your work, and I have urged
him to give us lectures this year on botany. He
has promised to do so, and I congratulate my
pupils at the Hospital on his compliance.   I
question if you could find a better comparative
anatomist, and he is a pearl worth your picking
up.   I assisted in drawing M. Délambre from
his retreat, and I beg you to help me in taking
M. Cuvier from his, for he is made for science
and the world.'   Such were the words of M.
Tessier; and I may be pardoned for introducing
them here, as they do more honour to our own
species than the history of great battles and
conquests."

M. Cuvier's grave and frequently absent air
has been repeatedly mistaken for an excess of
reserve and coldness, and thus it was often impossible for a mere casual observer to form a
correct judgment of the high degree of benevolence which he evinced to all who required

his assistance, the indulgence with which he viewed the follies of youth, and, in fact, the errors of all mankind. I may go still farther, and say the mirth which, before the death of his daughter, was to be traced in the laugh which seemed to proceed from the very heart. No one enjoyed a ludicrous circumstance more than he did; no one was happier at the performance of a comedy; for, when I was living in Paris, a ridiculous afterpiece was frequently represented on the stage, called " Le Voyage à Dieppe," in which the professors of the Jardin des Plantes were brought forward in the most amusing way possible; and such was M. Cuvier's uncontrollable risibility at its performance one evening, that the people in the pit several times called out to him to be quiet.

The nerves of M. Cuvier were particularly irritable by nature, and frequently betrayed him into expressions of impatience, for which no one could be more sorry than himself, the causes of which were immediately forgotten; and the caresses and kindnesses which were afterwards bestowed, seldom seemed to him to speak sufficiently the strength of his feelings at his own imperfection. Any thing wrong at table, to be kept waiting, a trifling act of disobedience,

roused him into demonstrations of anger wnich
were occasionally more violent than necessary,
but which it would have been impossible to
trace to any selfish feeling; even the loss of his
own time was the loss of that which was the property of others; and, where his mere personal
inconvenience was concerned, he was seldom
known to give way to these impetuous expressions. It was almost amusing to see the perfect
coolness with which the servants, more especially
about his person, occasionally obeyed his orders,
or replied to his injunctions without exciting
a hasty word from him. His impatience, however, was not confined to little annoyances; but
if he expected any thing, or any body, he
scarcely rested till the arrival took place. If he
had work men employed for him, the alteration
was done in his imagination as soon as com-
manded; and thus in advance himself, he unceasingly inspected their labours, and hastened
them in their tasks. He would walk along the
scene of operation, exclaiming every instant,
" Dépêchez vous, donc," (make haste, then,) and
impeding all celerity by the rapidity of his orders.
Perhaps, at the moment of pasting the paper on
the walls, he brought in a pile of engravings to
be put on afterwards, and which, in fact, were

often nailed up before the paste was dry. But although he was perfectly happy while thus engaged, he could not be alone, and, fetching his daughter-in-law back as often as she escaped from him, he associated her in all his contrivances. On unpacking a portrait of this ever ready companion by Sir Thomas Lawrence, and sent over from England, he happened to be present; and, in order to prevent him from seeing it by degrees, and so destroying the effect, she was obliged to hold her hands over his eyes, or he could not have resisted the desire to look. When he sent a commission to this country, every succeeding letter brought an enquiry as to its execution, or a recommendation to use zealous despatch. I must add, that the thanks were as often repeated as the injunctions. It is, perhaps, a curious inconsistency, that a man who submitted to such tedious and minute labour as he had all his life undergone, should be thus impatient when the activity of others was in question; but it must be recollected, that he found very few who would work as he did; and that, while so working, his mind was absorbed by every step which was taken to ensure the wished-for result, and had no time to bound over the space between thought and execution. " M. Cuvier possessed

in the highest degree, that patience which has
been said to be always necessary for the discovery
of some important truth, and which, according
to Buffon, and according to M. Cuvier himself,
constitutes the genius of a well-ordered mind.
No labour, however minute, irritated him when
he believed it to be requisite for the attainment
of his object; and this patience was really a virtue
in that man, whose blood would boil at a false
reasoning, or a sophism,—who could not listen to
a few pages of a book that taught nothing, or a
work that bore the marks of prejudice or passion,
without feeling the greatest irritation; and so
far did he carry his patient investigation, that he
even examined the least details of those elementary books which were to further instruction, and
directed the construction of several of the geo-
graphical maps of M. Selves, himself colouring
the models." *

In person M. Cuvier was moderately tall, and
in youth slight; but the sedentary nature of his
life had induced corpulence in his later years, and
his extreme near-sightedness brought on a slight
stoop in the shoulders. His hair had been light
in colour, and to the last flowed in the most
picturesque curls, over one of the finest heads

* Laurillard.

that ever was seen. The immense portion of
brain in that head was remarked by Messrs. Gall
and Spurzheim, as beyond all that they had ever
beheld; an opinion which was confirmed after
death. His features were remarkably regular
and handsome, the nose aquiline, the mouth full
of benevolence, the forehead most ample ; but it
is impossible for any description to do justice to
his eyes. They at once combined intellect, vivacity, archness, and sweetness; and long before
we lost him, I used to watch their elevated expression with a sort of fearfulness, for it did not
belong to this world. There are many portraits
published of M. Cuvier, formed of various materials; but, with the exception of the medallion
of M. Bovy, the copper medal, the plaster bust,
the lithographic print by M. le Meunier, and the
oil painting by Mr. Pickersgill, they scarcely
convey any just idea of M. Cuvier's expression :
in fact, some of the prints are positive caricatures. The bronze bust, modelled, and so
handsomely presented to the Royal Society of
London, by the celebrated sculptor, M. David,
was made from a cast taken after death. All
praise must be given to this bust as a work of
art; but it is very evident that M. David's
feelings, as an artist, were most susceptible to

the classic beauties of M. Cuvier's head and features (which, in fact, were remarkable), and, by
dwelling with too much stress on these, he has
lost sight of the benignity of the countenance.*
The bronze bas relief, taken from the bust, of
course possesses the same faults. Mr. Pickersgill's portrait is decidedly the most perfect of
all : it is there possible to form a correct notion
of the sharply defined features ; the eyes that so
well spoke the serious and great thoughts within,
that rose above this world ; the mouth, and the
carriage of the head. To use Mr. Pickersgill's
own words, he tried " to catch the essence of
the man," and his skill has proved adequate to
the great task before him.†

* Since writing the above, I have seen the bust worked in
French marble, after the same model, and given to Madame
Cuvier by this generous and public-spirited artist. It is an
improvement on that cast in bronze, and now stands on a pe-
destal in the room, and on the very spot where the mortal
remains of the great original were laid till they were removed
for ever.

† I cannot quit this subject entirely, without placing
Mr. Pickersgill in a still more admirable light than in his profession of artist. Feeling the value of the above-mentioned
portrait, after she lost her noble husband, Madame Cuvier was
naturally desirous of possessing a copy of it, from the hands
which had so well known how to execute the first. I was requested to negotiate concerning the possibility of sacrifice of
time, price, &c.; and the result was, that Mr. Pickersgill himself made the wished-for copy, which was not inferior to the

That love of order which so prevailed in great things was, by M. Cuvier, carried even into the minutiæ of life. His dissecting dress, it is true, was not of brilliant appearance, but it was adapted to the occasion; in this he would frequently walk about early in the summer mornings, in the open air, or pace up and down the galleries of anatomy, but on all other occasions his toilette was adjusted with care; he himself designed the patterns for the embroidery of his Court and Institute coats, invented all the costumes of the University, and drew the model for the uniform of the council, which drawing accompanied the decree by which it was esta-

---

previous likeness, and presented it to the Baroness, saying, that his services could be no affair of money between him and the widow of the great Cuvier. The sad delight with which the survivors accepted this generous gift was the highest reward which the donor could receive, and is best pourtrayed by their own expressions to me;—" C'est lui; c'est sa pensée, noble, pure, élevée, et souvent mélancolique, quoique toujours bienveillante et calme, comme la vraie bonté. C'est son âme dans ses yeux. C'est le grand homme, passant sur la terre, et sachant qu'il y a quelque chose au-delà." (" It is he; it is his noble, pure, elevated mind, often melancholy, though always benevolent and calm, like real goodness. It is his soul in his eyes. It is the great man passing over this earth, and knowing there is something beyond.") I may be forgiven for relating these anecdotes of the private feelings of the living, when it is considered how refreshing and useful it is to meet with such actions in this world of self-interest.

blished.   I was very anxious to see him in his
University robes, and having mentioned my wish,
he came into the room where I was sitting,
when decked in all the paraphernalia for a grand
meeting. The long, flowing gown of rich, violet-
coloured velvet, bordered with ermine, added to
his height, and concealed the corpulence of his
figure ; the cap, of the same materials, could not
confine his curls ; and, brilliant with his ribands
and his orders, the outward appearance fully
accorded with the internal man.   His refined
taste was often manifested in the buildings of
the Jardin, made according to his direction, and
was extended to the minutest details.   The menagerie for the wild beasts is classically beautiful, and was built entirely after his designs and
under his inspection, while he held the annual office of director.   The new wing of the Museum,
which joins the Corps de Garde, was also added
by his orders during one of these directorships.

The manners of M. Cuvier, by their dignity,
resembled the ancient deportment of French
people, divested of its extreme ceremony ; for,
accustomed to mingle with the highest of all
classes and countries, and naturally desirous of
paying a just tribute of respect and good-will to
every body, he was likely to be generally po-
lished and courteous, though in company, at the

houses of others, sometimes stately. That state-
liness was often deemed stiffness; and it must
not be denied that real stiffness was assumed on
some rare but necessary occasions. Frequently,
however, I believe that it arose from timidity;
for it wore off the instant he saw any one inclined to lay aside the restraint which his presence very often, and most needlessly, imposed.
On the contrary, when he saw people afraid of
him, he fancied he must have caused it by something on his side; and thus a counter-reserve was
produced, that seldom ceased with either party.
To the young however, he was universally encouraging, and they could not more entirely win
his heart than by talking, in his presence, in
their naturally open manner. Towards females
he was particularly kind and attentive, distin-
guishing all whom he thought worthy of more
than the general respect he paid to the sex,
even appealing to them on various occasions, delighting in their sensible remarks, and listening
to their anecdotes with the greatest interest. His
attentions to his guests, either when visitors for
a few hours or a few weeks, were surprisingly
thoughtful; if he could, he would have prevented their wishes, enquiring if they had all
they required in their own rooms, summoning

them to the drawing-room, if, by chance, any
one arrived whom he thought they would like to
see, expressly inviting those to his house who
had excited either their curiosity or interest, and
devising every thing he could think of for their
enjoyment or entertainment. At the time when
Paris was half mad about the Greeks, he suddenly re-appeared, after he had taken leave of
us, with a beautiful Greek boy, the son of Colo-
cotroni, whom he had accidentally met as he
quitted the Jardin ; but, fancying that we should
like to be acquainted with this intelligent, animated child, he took the trouble of coming back
on purpose to present him to us.   He frequently
walked, or rode home in a cabriolet, in order to
lend his carriage to the ladies of his house ; if a
wish was expressed to see a scarce book that his
own immense library did not contain, he would
bring it home from the Institute for inspection ;
and, while carrying on the most important duties
of the savant and the legislator, he yet found
time to think of others and their trifling desires.
Now and then, when the summer lessened some
of his heavy public duties, he would take a walk
with us; and no schoolboy, with permission to
go out of bounds, could set off with more delight than we all did. Sometimes he would con-

fine himself to the Jardin; and in one of these more limited excursions he was attracted by the brilliant appearance of the Coreopsis tinctoria, which was then new in France, and which he saw for the first time during this ramble. He in vain enquired the name of us, and we continued our walk. On returning to the house, he quitted us at the door, and, in about half an hour, he re-appeared, and, stopping, for an instant, as he descended to his carriage, he said, " Ladies, I have been to M. Deleuze (a learned botanist of the Jardin), and ascertained the name of the flower:" he then gave it us, genus, species, country, and the reason for its appellation, and, making his bow, retired, perfectly happy at the knowledge he had acquired and imparted. As in this trifling circumstance, so was it in all things; he never hesitated saying when he did not know; he never rested till he did know, if the means of acquiring the information were within his reach; and, once known, he was most willing to impart it to those who wished to learn. The facility with which he placed knowledge in the reach of others was one of the most precious gifts with which Heaven had endowed him; for half the value of a brilliant or an useful idea is lost, unless we have the power of communicating

it as it appears to ourselves. Sometimes he
would enliven the evening by proposing a party
to eat ices at one of the famous cafés; and, on
one occasion, he insisted on showing me, as an
Englishwoman, how happy the lower classes of
French are on their fête days; and, passing the
barrier close to the Jardin des Plantes, he led
us among the guinguettes * outside, where the
people were dancing and singing, and making
merry. He delighted in their mirth, stopped to
witness it, and, several times turning round to
me, asked me if the English knew any thing of
such light-hearted enjoyment. It is said of some
celebrated person, that no one could take shelter
from the rain with him, under a shed, for a
quarter of an hour, without deriving some information from his discourse. This observation
may be equally applied to M. Cuvier; for after
these little excursions, intended solely for diver-
sion, it was frequently a matter of surprise to
find that something had been learned, either by
way of history, character, language, or moral

* Many of these guinguettes consist of nothing but a mere
shed, with a little space in front, where the guests sit and
drink weak wine (vin ordinaire), sugar and water, lemonade,
&c., dance, sing, and play at dominos. They are generally
placed outside the barriers to avoid the duty paid on pro-
visions of every kind as they enter Paris.

conduct; so elevated, yet so fascinating, was the
tone of his unrestrained conversation.

M. Cuvier's hours of audience generally took
place before and after breakfast, and he was accessible to every body; for he said, " when
people lived at such a distance as the Jardin des
Plantes, they had no right to send any one away
who came so far to request their advice or assistance." I have seen the young and the old,
the widow and the orphan, the poor and the
rich, throng his door, all in the security of being
well received. I met an unhappy woman one
morning, crying as she came down stairs; and
on asking her what was the matter, she replied,
" It is not M. Cuvier who has made me cry;
but it is because even he cannot help me that I
am in such trouble;" evidently thinking that, if
he could not serve her, she had no hope. The
meal-times were always anticipated by his family
and friends with the greatest pleasure; for then
it was that questions were asked, and histories
related on all sides. As if knowing the few opportunities there were of conversing with him,
he suffered himself to be constantly interrupted,
and never hesitated giving the desired information concerning public or private circumstances; and frequently, when the former were

not immediately comprehended, he would set forth the subject in all its bearings, till it was perfectly understood. The breakfast took place generally at ten; but M. Cuvier had almost always risen at seven, or even before that time, had prepared his papers for the day, had arranged the occupations of his assistants, and had received most of his visitors. Some intimate friends frequently called on him at this hour, because they were sure to find him at home. His usual practice was to read the newspapers as he ate his breakfast, or look over the books for the use of the primary schools, sent for his inspection. Still, if one of the family were missing, he would enquire for the absent person with the utmost solicitude; and even if the conversation were unusually animated, he insisted upon knowing the whole, though he seldom raised his eyes from the paper. After breakfast was finished, he dressed, and then came the routine of his numerous occupations; and when it was his turn to be Director of the Jardin, before going to the Council, &c., he would take his way, amid the trees, to the Museum of Natural History, followed by secretaries, aide-naturalists, students, &c., bearing the treasures which had just been finished in the stuffing laboratories, and which were arranged in their respective cases under his superintendence. His carriage was generally punctual to the moment appointed, and no one was allowed to keep him waiting; and, in fact, no one would do so, if possible to avoid it, for it vexed him exceedingly; though I used to think sometimes that I saw a faint smile on his countenance, when we flew down stairs, our gloves in our hands, and our shawls streaming after us. The instant he had given his orders, he would thrust himself into a corner, and resume his reading or writing, suffering us to talk as much as we pleased. Many of his most brilliant memoirs were finished as he thus rode through the streets of Paris; and he had a lamp fixed to the back of his carriage, that he might read on his return home at night from his visits; but he found it so distressing to his eyes, that he could not long make use of it. All others, however, were delighted at the disappointment, because he was by it cheated into a few more moments of repose.

Privileged as Mr. Bowdich and myself were to inspect the vast treasures in his collections, and in his library, at our leisure, we yet found it much more agreeable to take the books home with us; frequently we required the very volume to which he had been referring before his departure, and which was generally left open upon his table, to be again used on his return; for he had the happy faculty of resuming his subject at any moment, in any place, and at any part, even in the middle of a sentence. Waiting, then, till his carriage was driven from the door, bearing him away for several hours to his administrative duties, we went up to his room, took possession of the book, and enquiring the hour of his return, fled back with it five minutes before it was wanted. To be sure, in consequence of our having been a little too late on one or two occasions, — a circumstance which he bore with surprising good humour, — we used occasionally to see some of his household arrive at our hotel, in breathless haste, to enquire for a volume which had long been missing. Generally speaking, we were innocent of the misdemeanor; but such was his indulgent goodness to us, that he not only facilitated every desire, every endeavour to obtain improvement, but even allowed us to publish, for the first time, some of his own drawings of Mollusca. He had no idea of exclusion towards any one who he thought would make a proper use of the materials he could furnish; so that we had only to ask, and orders were given to the keepers of the galleries to take out of the cases any object which was needed for our closer examination.*

Before dinner, M. Cuvier would occasionally give a few minutes to his family, by joining the assembled party in Mme. Cuvier's room. On the sound, " Madame est servie," he would offer his arm to his wife, and leading her to her seat, all gathered round them both at this once happy table. M. Frederic Cuvier, his son, and very often one or two intimate friends who came by chance, would increase the circle, and the most

* Though perhaps somewhat foreign to my subject, I cannot forbear making use of the first opportunity afforded me of expressing my gratitude to many connected with this vast and magnificent establishment. M. Desfontaines, M. de Jussieu, M. Brongniart, M. Geoffroy St. Hilaire, M. Frederic Cuvier, M. Chevreuil, M. Valenciennes; M. Deleuze, and M. Laurillard, thank God! still live to receive this public testimony of my sense of their kindness. M. Haüy, M. Latreille, M. Thouin, M. Royer, M. Dufresne, W. Vanspaendonck, M. Lucas, have been called to another world, where human feelings are of no avail. Our pass-word for every thing was, " de la Maison Cuvier;" and night and day we wandered about this little world as if we had been among its permanent inhabitants. Great have been the changes since then; and now the master spirit of this beautiful abode is no more, I shall never look on it again, and fancy that it has retained its perfection. During my late visit, not even the subordinate employés whom I had known in former times, but, after their respectful greeting, lamented the death of their great patron, in words that betokened the sincerest grief.

delightful conversation would ensue. On proceeding to the drawing-room, M. Cuvier would
occasionally gratify those present by an hour's
stay among them before he retired to his occupations, or paid his visits. Occasionally he would
bring forth some old book he had picked up at a
stall on one of the Quais, and boasting of his
bargain, read some passages; or, bidding some
one read to him, he compared different editions. At a more recent period, if he had any
of M. Champollion's letters from Egypt, he
would station us at different tables, with volumes
of the great work on Egypt, and verify the descriptions of the antiquary step by step. He
never was weary of research; though, it must be
owned, we occasionally wished for the sound of
the carriage wheels, to interrupt our employment.
He never suffered people to be idle in his house;
and no sooner did friends station themselves
among the family for a time, but he would come
into their rooms with folios and paper in his
hand, and set them to trace plates for him; and
seldom forgot, on his return home from his duties
abroad, to enquire how much had been done. To
be sure, it was a pleasure to work for him, he
was so grateful for the service, and so happy
when the task was properly completed. His

thirst for knowledge took an unbounded range,
and the inventions and enterprises of other
countries were as interesting to him as those of
his own. Every letter to me, at the time that
the accidents happened to the tunnel under the
Thames, contained enquiries concerning it: the
steam carriages, railroads, suspension bridges,
and public institutions, were all subjects of correspondence: he read, or made others read to
him, all the attempts that had previously taken
place to perfect the same undertaking; and when
a person from the country in which the scheme
was going forward came to see him, he was prepared to converse with the stranger as one deeply
learned in the matter. He was one day talking
to a gentleman high in office at one of our
national establishments; and after mentioning the
expenses of the Museum, &c. at the Jardin des
Plantes, he to the great surprise of his companion, stated to a fraction the former, and actual
costs of the British Museum. He could not
bear to be inactive for an instant; and once, while
sitting for a portrait, which was to face the
quarto edition of his " Discours sur les Revolutions du Globe," Mlle. Duvaucel read to him
the " Fortunes of Nigel." He had a map of
London at his elbow, which the artist allowed

him occasionally to consult; and the Latin of
King James often excited a smile, which was a
desirable expression for the painter; but unhap-
pily the engraver was not a faithful copyist, and
this published portrait is anything but a resemblance.

A change of occupation was a relaxation to
M. Cuvier; perhaps the greatest of all was conversation; but there was yet a third, which was
to throw himself on a sofa, hide his eyes from
the light, and listen to the readings of his wife
and daughters, and, occasionally, that of M.
Laurillard. By these nightly readings, for they
only took place when he could not work any
longer, he became acquainted with the literature
of the whole civilised world; and no one was
better able to appreciate it, for he looked on it
as a picture of the human mind, and judged by
it of the state of civilisation in the country
where the various works were published. He
frequently thus renewed his acquaintance with
books read long before, in order to mark the
changes which had taken place in the lapse of
years; and the number of volumes perused in
this way was immense, though the reading
seldom or never lasted more than two hours.
There was yet another advantage which attended

this manner of closing the day by such a rational amusement: it served to quiet his mind,
which had often been previously excited; and
ensured him that undisturbed repose, which
fitted him for succeeding labours, and which his
appearance the next morning generally indicated
that he had enjoyed. Could that man's slum-
bers be otherwise than sweet, who had passed
every moment in the fulfilment of the most important duties of life? The services thus rendered to M. Cuvier were often returned by him
in kind; for if any member of his family was ill,
he would take his books and his newspapers to
the bed-side, and read aloud by the hour together. He never slept except at night; and I
never heard of any one surprising him in such a
state of inaction at any hour in the day, in his
house or carriage, whatever might have been the
fatigue he had undergone.

No one was ever more sensible to kindness
than M. Cuvier, and the slightest services always
received acknowledgments beyond their value;
it is not surprising, then, that in the same character there should be an equal sensibility to ingratitude. To find any one thus return the
affectionate cares he had bestowed, was a real
affliction; and as an instance, among several

others, I recollect that, during one of my visits to his house, he appeared most unusually sad, and all the efforts to amuse him were repaid by a mournful smile. All his family were aware that no calamity could have produced this, and guessed it was some trouble connected with others, into which they had, perhaps, no right to enquire; and they were not wrong in their conjectures. Walking home one evening quietly with his daughter-in-law, in reply to her remarks upon his dejection, he confessed that a favourite friend and pupil had, from motives of self-interest, publicly sided with his enemies, and it was an affliction to which he could not easily reconcile himself.

The benevolence of M. Cuvier was evinced in every form by which it could be serviceable to others; and students themselves have told me, that he has found them out in their retreats, where advice, protection, and pecuniary assistance were all freely bestowed. Frequently did his friends tax him with his generosity, as a sort of imprudence; but his reply would be,—" Do not scold me, I will not buy so many books this year." Many anecdotes have been told me of his purse being made a resource, not only for the advantage of science, but for those who had fled to France to avoid ruin in their own country; but even my anxiety to make known all M. Cuvier's good qualities ought not to interfere with the sacredness of private misfortune. In his endeavours to do good, he was always most ably seconded by the females of his family, whose active benevolence has called upon them many a blessing from the hearts they have cheered by their kindness and bounty.

A very remarkable and a very prominent feature in M. Cuvier's character, was a decided aversion to ridicule or severity when speaking of others: he not only wholly abstained from satire himself, but wholly discouraged it in those around him, whoever they might be; and was never for one instant cheated into a toleration of it, however brilliant the wit, or however droll the light in which it was placed; and the only sharpness of expression which he allowed to himself, was a rebuke to those who indulged in sarcasm. On hearing me repeat some malicious observations made by a person celebrated for his wit and talent — not being aware of the hidden meaning of the words I quoted, and having been very much amused with the conversation— M. Cuvier instantly assumed a gravity and seriousness which almost alarmed me, and then solemnly bade me beware of the false colouring which I was but too apt to receive from the person in question; but fearing I should feel hurt, he instantly resumed his kindness of manner, and lamented that the real goodness of heart, the great abilities, and power of divesting himself of partiality, in my friend, should so often be obscured by the desire of saying what was clever or brilliant.

Two other great traits — perhaps, I ought rather to call them perfections — belonging to M. Cuvier, were, a total absence of all self-conceit and all resentment, both of which led to a remarkable uniformity and kindness in performing the duties of social life. That he had preferences, and that these preferences were sometimes formed from the first interview, was true; and few people possessed of quick and ardent feelings can avoid these sudden impressions; but a contrary feeling led him merely to avoid intercourse, and did not, in any manner, extend towards the real welfare of the individual. Even the annoyances and disappointments he met with in his public career left not one grain of bitterness in his soul; and he generally laid the fault to the ignorance, rather than the bad feeling, of the offenders; saying of them, — " They are more

to be pitied than blamed, for they know not
what they do." No one knew better how to
soften a refusal; and, whatever might be his
reasons, he took care that his opinions should
not wound the feelings of any applicant for his
favours. During one of my visits to his house,
a gentleman, anxious to obtain the vote of M.
Cuvier, as serviceable in procuring a public employment, applied to me to intercede with my
noble host. I felt that I had no right to do so,
and mentioned my dilemma to Madame Cuvier,
at the same time expressing my vexation, that such
advantage should have been taken of my intimacy.
This being repeated to M. Cuvier, he laughed
at the scruples which had withheld me from conversing with himself on the subject, and then
desired me to reply to the applicant, that he
never suffered the ladies of his family to interfere
in such matters. When I left the room in
order to do this, he called me back, as if a sudden thought had struck him; and he added, —
" Tell your friend, if he wishes to see me, or
ask my advice, I shall be happy to receive him
at ———— ;" evidently wishing to save me from
the pain of an abrupt refusal to one whom I
might esteem.

The soirées of Baron Cuvier, which took

place every Saturday evening, and were sometimes preceded by a party, were the most brilliant and the most interesting in Paris. There,
passed in review, the learned, the talented, of
every nation, of every age, and of each sex; all
systems, all opinions, were received; the more
numerous the circle, the more delighted was the
master of the house to mingle in it, encouraging,
amusing, welcoming every body, paying the utmost respect to those really worthy of distinction, drawing forth the young and bashful, and
striving to make all appreciated according to their
deserts. Nothing was banished from this circle
but envy, jealousy, and scandal; and this saloon
might be compared to all Europe; and not till
the guest had repassed the Rue de Tournon, or
" the Seine, could he again fancy himself in the
capricious capital of fashion, or time-serving
show." It was at once to see intellect in all its
splendour; and the stranger was astonished to
find himself conversing, without restraint, without
ceremony, with, or in presence of, the leading
stars of Europe: princes, peers, diplomatists,
savants, and the great host himself, now receiving
these, and now the young students from the fifth
pair of stairs in a neighbouring hotel, with equal
urbanity. No matter to him in which way

they had directed their talents, what was their
fortune, what was their family; and wholly
free from national jealousy, he alike respected
all that were worthy of admiration. He asked
questions from a desire to gain information,
as if he too were a student; he was delighted
when he found a Scotchman who spoke Celtic;
he questioned all concerning their national institutions and customs; he conversed with an
English lawyer as if he had learned the profession in England; he knew the progress of public
education in every quarter of the globe; he
asked the traveller an infinity of things, well
knowing to what part of the world he had
directed his steps; and seeming to think that
every one was born to afford instruction in some
way or other, he elicited information from the
humblest individual, who was frequently astonished at his interest in what appeared so familiar to himself. One thing used particularly to
annoy him; which was, to find an Englishman
who could not speak French. It gave him a restraint of which many have complained, but
which, on these occasions, solely arose from a feeling of awkwardness on his own part at not being
able to converse with his foreign guest. No one
ever rendered greater justice to the merit of his
predecessors or contemporaries than M. Cuvier.

" Half a century," he said, " had sufficed for a
complete metamorphosis in science; and it is
very probable that, in a similar space of time, we
also shall have become antient to a future generation. These motives ought never to suffer us
to forget the respectful gratitude we owe to those
who have preceded us, or to repulse, without
examination, the ideas of youth; which, if just,
will prevail, whatever obstacles the present age
may throw in their way." This was a delightful
manner of satisfying every body with himself: the
naturalist, from a remote province, or perhaps
from a colony at the other end of the world, was
no longer ashamed to think that he had not
kept pace with the march of science in the capital,
and had been poring over obsolete systems; and
the young student, fresh from the Universities,
was not afraid to utter the objections, the falla-
cies, or the inaccuracies, he fancied he had
detected in his perusal of more recent authors.

The repast which closed these evening enter-
tainments was served in the dining-room, and,
certainly, at the most delightful tea-table in the
world. A select few only would stay, though
M. Cuvier sometimes pressed into the service
more than could be well accommodated; and
while the tea, the fruit, and refreshments of

various kinds were passing round, the conversation passed brilliantly with them. Descriptions of rarities were given, travellers wonders
related, works of art criticised, and anecdotes
told; when, reserving himself till the last,
M. Cuvier would narrate something which
crowned the whole; and all around were either
struck with the complete change given to the
train of thought, or were forced to join in a general shout of laughter. One evening, the various signs placed over the shop doors in Paris
were discussed; their origin, their uses, were
described; and then came the things themselves.
Of course, the most absurd were chosen; and,
last of all, M. Cuvier said that he knew of a
bootmaker who had caused a large and ferocious
looking lion to be painted, in the act of tearing
a boot to pieces with his teeth. This was put
over his door, with the motto, " On peut me
déchirer, mais jamais me découdre." *   I was in
Paris when the celebrated picture, painted by
Girodet, of Pygmalion and the Statue, was ex-
hibiting at the Louvre. It caused a general
sensation; epigrams, impromptus, were made
upon it without end; wreaths of flowers, and

---

* " I may be torn, but never unsewn."

crowns of bays, were hung upon it; so that it became an universal theme of conversation. Among other topics, it was one evening introduced at M. Cuvier's; when M. Brongniart (the celebrated mineralogist, and director of the Royal Manufactory of China at Sévres), found fault with the flesh, which, he said, was too transparent; Baron de Humboldt (the learned Prussian traveller, who had lately been occupying himself with the chemical experiments of M. Gay-Lussac) objected to the general tone of the picture, which, he said, looked as if lighted up with modern gas; M. de Prony (one of the mathematical professors of the Ecole Polytechnique, and also director of the Ecole des Ponts et Chaussées *) found fault with the plinth of the statue; and many gave their opinion in the like manner, each pointing out the faults that had struck him in this celebrated performance; after which, M. Cuvier said that the thumb of Pygmalion was not properly drawn, and would require an additional joint to those given by nature, for it to appear in the position selected by the painter. Upon this, M. Biot (the mathematician and natural philosopher, who had re-

* A school resembling those for our civil engineers.

mained silent all the time,) with mock solemnity
summed up the whole, showing that every body
had been more or less influenced by his peculiar
vocation, or favourite pursuit; and concluded
by saying, that he had no doubt but that every
one of them, if they met Girodet the next day,
would congratulate him on the perfect picture
he had produced. On these evenings, one or
two old, or particularly cherished friends would
remain, talking after the rest had taken their departure; the hours passed, the clock would
strike two before the little coterie thought of
separating; and even then M. Cuvier would say,
" Nay, gentlemen, do not be in such a hurry, it
is quite early."

But I am now speaking of that period which
preceded the death of the angel Clementine, so
named after the dear and excellent mother, who
had so well guided the earliest youth of her
father. This pure creature was so good herself,
that she never suspected evil in others, and was
the light of every body's existence in this hal-
lowed circle. Her likeness to M. Cuvier was
very striking; and though her eyes and hair
were of a darker and a different shade, his every
feature could be traced in her countenance,
softened into feminine beauty. Her talents, her

x

acquirements, her modest opinion of herself, her
sound judgment, her active charity, her extreme
piety, seemed to mark her as a being who could
not long remain in this world of sin: she died
of rapid consumption, which disease, though
probably, long engendered in her constitution,
which had already given one or two alarms,
and probably made hidden progress, only manifested itself in its decided form six weeks before
her death, amid the joyful preparations for her
marriage. From this moment a mournful change
took place in every arrangement; the broken-
hearted mother was long, very long, unable to
receive company, never again to mingle in it
abroad; and the unceasing and heroic efforts of
her surviving daughter, and the affectionate
cares of her husband, failed to rouse her. At
length, occasional society at his own house became absolutely necessary to M. Cuvier, and the
good wife consented to that which the good
mother had refused; the saloon remained closed
in which she had seen the perfection of mortal
loveliness breathe her last, and one of the libraries was opened to company. A few old friends
alone took immediate advantage of the permission to resume their visits; these, in time,
brought others; but the change had come; and

to those who had known Clementine, the soirées
were stripped of one of their principal charms.
In vain did M. Cuvier exert himself more than
ever to welcome his guests; vain was the conversation of his daughter-in-law, the most fascinating and brilliant that perhaps ever fell
from the lips of woman; there sat the dejected
mother, evidently making an effort over herself,
her thoughts but too plainly in another sphere;
and the cause of her abstraction was whispered
to strangers, with mournful looks and faltering
tongues, by those who had beheld the being that
had filled up the vacancy. With a violent effort,
that closed saloon was once more opened to M.
Cuvier's friends; but it seemed to be only the
preparation for the dying breath of the parent.
That saloon is now always open, and the bereaved widow and her devoted child always inhabit it, surrounded by the portraits of those
they loved, clinging to the shadows and recollections of those that are gone, and living in the
past, as the sole source of their melancholy enjoyment.

After the death of his own daughter, M. Cuvier became, if possible, more than ever attached
to Mademoiselle Duvaucel. He had never made
any difference in his conduct towards her and

Mademoiselle Cuvier; but the loss of the latter
necessarily increased his reliance on her cares,
and an anxiety was added to his affection, which
manifested itself on all occasions. If she were
ill, ten times in the course of the day would he
mount up stairs to enquire at her bedside how
she felt; if she coughed it seemed to give a
pang to his very heart; and, on her part, could her
redoubled devotion towards him and her mother
have filled up the void, their great loss would
have been repaired.

In 1830, as we have already seen, M. Cuvier
paid his last visit to England, in which journey
he was accompanied by Mlle. Duvaucel, who
was willingly spared by her mother; for so
fondly had these two beings watched over him,
that he almost required the one or the other to
be constantly with him. This visit, happening
as it did during the period of the last revolution,
caused several reports in this country of their
having fled to avoid danger. Hearing these
surmises whispered about the hotel where they
resided, M. Cuvier's faithful valet ventured to
repeat them, and asked his master if he were
really ignorant of what was about to take place.
" Do you think, Lombard," replied M. Cuvier,
mildly, " that if I had been aware of that which

was about to happen, I should have left Madame
Cuvier ?" To those who knew the man, this
answer was the best refutation to such suppositions. The fact was, that the opportunities of
absenting himself were rare, and feeling the
necessity of coming to England for scientific
purposes, more especially connected with his
great work on fishes ; feeling, also, that a change
was required by his constitution, so overcharged
with mental labour, a mere apprehension was not
likely to deter him from a project which had
been delayed in its execution by a concurrence
of untoward circumstances. Till M. Arago was
elected in the place of Baron Fourrier, M. Cuvier
could not quit his Secretaryship of the Academy
of Sciences, the duties of which were doubled
by the death of the latter ; and, further than that,
it was requisite for him to read his admirable
éloges on Sir Humphry Davy and M. Vauquelin,
at the next general meeting of the Institute, and
the postponement of that meeting threw another
obstacle in the way of his immediate departure.
I have already mentioned how deceived he was
by the apparent tranquillity of Paris on the
morning in which he left it, and how he was induced to proceed, even after he had determined
to return from Calais.

As I have here spoken of the meeting of the
Academies on the 26th of July, I will stop to
correct an error which has obtained much circulation in England.   A personal quarrel is said to
have taken place on that day, before the meeting,
between M. Cuvier and M. Arago, in which the
former was, with difficulty, prevented from drawing his sword.   The only foundation for this
report was, that M. Arago was obliged on this
occasion to read an éloge on M. Fresnel, in
which he had introduced a very violent paragraph
against the Clermont-Tonnerre ministry, which
paragraph might easily have been converted into
a marked reference to the then existing government.   M. Cuvier suggested to M. Arago that
it would be more wise and prudent to leave out
this part of the éloge, and at such a moment to
avoid all causes of excitation.   He gave his advice in the most friendly manner; but, as M.
Arago defended his paragraph with considerable
warmth, M. Cuvier ceased to urge the matter.
After this, the two secretaries appeared together
before the public assembly, in the Hall of the
Institute, and when the ceremony was concluded
they dined together at M. Cuvier's house, and
passed the evening most amicably in each other's
society, without an idea that their mere differ-

ence of opinion would cross the Channel in the shape of a dreadful and almost murderous quarrel.

The first intention of the travellers was to proceed by way of Dover; but, to please Mademoiselle Duvaucel, M. Cuvier ascended the river, and landed at the Tower stairs. Often did he congratulate himself, afterwards, on this compliance, which afforded him a view of the banks of the Thames, and the thousands of vessels which float on its surface, and of which no foreigner can possibly form an idea without actual inspection. The object of one of M. Cuvier's first walks, after his arrival in London, was to see all the new caricatures contained in our shop windows; for he was a warm admirer of our performances in this art, and already possessed a voluminous collection of the best which had appeared. They afforded him more than mere amusement, for he considered them as curious documents of the moral and political history of certain periods; and often, in the midst of a serious conversation concerning the events of our own times, or those immediately preceding us, he would cite various circumstances which had been stamped upon his recollection by the sight of an English caricature. During

the fortnight he was in London, he was in incessant motion; but his anxiety respecting public
events embittered all his enjoyment. An acci-
dental circumstance delayed one of Madame
Cuvier's daily epistles, and he scarcely rested
during these hours of expectation. One morning, however, he flew into the room where Mademoiselle Duvaucel was with me, preparing to
go out, entered without the slightest ceremony,
embraced us both, and exclaimed, " I have
heard from my wife;" then, reading the letter,
he asked us if we were not as happy as himself;
and taking an affectionate leave, as if his heart
was quite full, he hastened to an appointment at
the British Museum. He made a great many
notes, and several drawings, while here, relative
to his Fossil Remains and Ichthyology, but contrived a few hours for visiting. The enlightened
and amiable Baron Seguier, the Consul-General
of France, was then living, and the little party
assembled several times at his house, where the
events then taking place in their own country
were constantly discussed, and where these able
men predicted much of that which has since
occurred. M. Cuvier went to Richmond also,
of which he had so frequently heard in terms of
praise : the day was rather stormy, but with in-

tervals of brightness, which added to the effect
of the scene; and he observed, that he could
not wonder, when he saw such a sky over such a
country, at the perfection to which the English
had carried their landscapes in water colours.
He had intended revisiting Oxford, and seeing
Cambridge, with the latter of which he was
only acquainted by report; but the curtailed period of his stay did not permit him to enjoy these
pleasures. Never, however, did any one profit
more entirely by every hour than he did. Accustomed to consider his insatiable desire to see
and know every thing as a virtue, he left no
means untried to satisfy his curiosity; he rose at
six, visited on foot various parts of London,
which he had never before seen, then returning
to breakfast, he entered his carriage with his
companion, and went to the Parks, the exhibitions, collections, &c. He was every where
pleased with the reception he met with, though
it was a matter of regret to the English that so
few persons chanced to be in the metropolis to
do him honour. One amusing mark of respect
was a source of great entertainment, and for its
drollery alone do I offer it to the reader. During the absence of his valet, M. Cuvier sent for
a barber to shave him. The operation being

finished, he offered to pay the requisite sum;
but the enlightened operator, who happened to
be a Gascon, bowed, and positively refused the
money, saying, with his comic accent, " he was
too much honoured, by having shaved the greatest man of the age, to accept any recompense."
Hardly suppressing a smile, M. Cuvier felt
bound to give him the honour to its full extent,
and engaged him to perform his function every
day while he remained in London. It is scarcely
necessary to add, that the barber, in a short
time, felt it a still higher duty to consult prudence rather than empty honour, and pocketed
the amount due for the exercise of his calling.

Although occasionally subject to slight ail-
ments, the health of M. Cuvier, generally speaking, was good, and his carriage was used by him
more as a saving of time than a matter of necessity; therefore the sudden summons he received
to quit his earthly labours, was an event for
which his friends and his country were not prepared. Never were his intellectual faculties more
brilliant; never was his great mind more fully possessed of that clearness, that comprehensiveness,
which so peculiarly marked it, than at the time
of his seizure. His life of temperance and recti-
tude, at the age of sixty-two, had preserved the

corporeal existence unimpaired, and also contributed to the perfection of his mental vigour; for
more than forty years he had been unremittingly
labouring to perfect his great views in science
and legislature; and concerning the former he
was about to give to the world the results of his
researches and reflections.   " His intention was
to review all his works, and put them on a foot-
ing with the last discoveries, and then to deduce
from them all the consequences, all the general
principles, which appeared to him to emanate
from such an assemblage of facts, though he did
not think it possible, in the present state of
human knowledge, to establish a general theory.
All his studies, all his meditations had convinced
him of the philosophical principle, that organised
beings exist for an end, for a special object; but
he did not admit any scientific theory, and with
all his energy maintained that it was not yet
possible for any to be formed." *   But even the
entire publication of these facts, of these deduc-
tions, was denied to us by the inscrutable ways of
the Almighty; perhaps we were not yet worthy
of penetrating so deeply into the mysteries of
creation as had been given to this one gigantic

* Laurillard.

intellect, and I dare not call the death of M.
Cuvier premature, when I think that by so doing
I should question the decrees of that Providence
to whom we owe the very existence of him
whom we deplore, by whom that life was lent to
us to increase our sense of his wisdom, and to
enlighten us by its example.

M. Cuvier had sought forgetfulness of the
storms that were passing without the walls of
his peaceful abode, in a greater devotion than
ever to his home pursuits; that is, he gave up
his evening visits, and the few relaxations he had
permitted himself to enjoy. The cholera raged
around him, and he saw those fall who were
younger and apparently stronger than himself;
those whom he loved, and those whose services
were essential to the state. Public disturbances
filled the streets of Paris, while pestilence stalked
through the multitude in every direction. Se-
cluding himself, then, entirely from society, except that of his family; after going through the
daily routine of his public duties, he returned to
his labours with an intenseness, which, added to
his share of the pervading gloom, was calculated
to injure the springs of life. No one, however,
could foresee its effects on his constitution;
and he himself said, " that he had never worked

with so much real enjoyment;" and he rapidly advanced, not only in the vast undertakings then begun, but in the preparations for others. On Tuesday, the 8th of May, he opened the third and concluding part of his course of lectures, at the Collège de France, on the History of Science, &c., by summing up all that had been previously said. He forcibly inveighed against that heresy in natural history, which derives every thing in this vast universe from one isolated and systematic thought, and shackles the future of science with the fallacious progress of the moment *: he pointed out what remained for him to say respecting the earth and its changes, and announced his intention of unfolding his own manner of viewing the present state of creation ; a sublime task, which was to lead us, independent of narrow systems, back to that Supreme Intelligence, which rules, enlightens, and vivifies, which gives to every creature the especial conditions of its existence, to that intelligence, in short, which reveals all, and which all reveals, which contains every thing,

* Alluding to the theory of unity of composition. This and the following citations are taken from a description of this admirable lecture, as noted by a distinguished auditor, the Baron de H———.

and which every thing contains. In the last
part of this discourse, there was a calmness, a
clearness of perception, an unaffected and unrestrained manifestation of the contemplative and
religious observer, which greatly added to its
force, and which involuntarily recalled that book
which speaks of the creation of the earth and
the human race. The similarity was avoided
rather than sought; it was not to be found in
the words, but the ideas; and at once flashed
across the minds of his auditors, when the great
professor declared, that each being contains in
itself an infinite variety, an admirable arrangement for the purposes for which it is intended;
that each being is good, perfect, and capable of
life, each according to its order and species, and
in its individuality.   In the whole of this lecture
there was an omnipresence of the Omnipotent
and Supreme Cause; the examination of the
visible world seemed to touch upon the invisible;
the search into creation, necessarily invoked
the presence of the Creator; it seemed as if the
veil were to be torn from before us, and science
was about to reveal eternal wisdom. Great then,
was the effect produced by the concluding sentences, which seemed to bear a prophetic sense,
and which were the last he ever addressed to his

audience. " These," said he, " will be the objects
of our future investigations, if time, health, and
strength, are given to me, to continue and to
finish them with you." Those who were versed in
human destiny, seemed to feel, that his sphere
of action was even then placed out of this world,
and that he had pronounced his farewell. So
near the great and awful tribunal, what other
words, what other thoughts than those contained
in this lecture, could have so plainly shown
the preparation already made for his journey
thither?

I am told that the profound emotion occasioned by this last discourse was universal, and
that few left the hall without an undefined feeling of sadness, and sentiments of reverence, far
beyond the power of expression. On the same
day, M. Cuvier, as usual, attended a council of
administration in the Jardin des Plantes, and
bestowed his last cares on that immense establishment, which owes so large a portion of its
treasures to his constant and active solicitude,
and to his extreme generosity. " By turns protected and protecting, M. Cuvier had there resisted the political vicissitudes which changed
all but this sacred asylum of men and things.
It would seem as if a special grace from Provi-

dence had suffered him to remain, during thirty-
eight years of revolution, in the same place, and
with the same occupations. The great mind, the
pure intention, the devoted and disinterested
heart, alone are suffered to effect such miracles."

In the evening of Tuesday, M. Cuvier felt
some pain and numbness in his right arm, which
was supposed to proceed from rheumatism. On
Wednesday, the 9th, he presided over the Committee of the Interior with his wonted activity.
At dinner that day, he felt some difficulty in
swallowing, and the numbness of his arm increased. Never can the look and the enquiry
he directed to his nephew, when he found that
bread would not pass down his throat, be forgotten; nor the self-possession with which he
said, as he sent his plate to Madame Cuvier,
"Then I must eat more soup," in order to quiet
the alarm visible on the countenances of those
present. M. Frédéric, the younger, sought medical advice; and an application of leeches was
made during the night, without producing any
amelioration. The next day (Thursday) both
arms were seized, and the paralysis of the pha-
rynx was complete. He was then bled, but
without any benefit, and from that moment he

seemed to be perfectly aware of what was to follow. He, with the most perfect calmness, ordered his will to be made; and in it evinced the tenderest solicitude for those whose cares and affection had embellished his life, and for those who had most aided him in his scientific labours. He could not sign it himself, but four witnesses attested the deed. He sent for that good M. Royer, who was so soon to follow him, to make a statement of the sums he had expended, out of his private fortune, on the alterations of the rooms behind his house, though the affliction of this Chef du Bureau d'Administration was so heavy as almost to disable him from doing his duty. M. Cuvier alone was tranquil; and, perfectly convinced that all human resource was vain, he yet, for the sake of the beloved objects who encircled him, submitted without impatience to every remedy that was suggested. The malady augmented during the night, and the most celebrated medical practitioners were called in: emetics were administered by means of a tube, but, like all other endeavours, they did not cause the least alteration. Friday was passed in various, but hopeless, attempts to mitigate the evil; and perhaps, they only increased the suffering of the patient. In the evening the paralysis attacked the legs; the night was restless and painful; the speech became affected, though it was perfectly to be understood. He pointed out the seat of his disorder, observing to those who could comprehend him, " Ce sont les nerfs de la volonté qui sont malades*;" alluding to the late beautiful discoveries of Sir Charles Bell and Scarpa, on the double system of spinal nerves†: he clearly and precisely indicated the changes of position which the parts of the limbs yet unparalysed rendered desirable; and he was moved from his own simple and comparatively small bed-room, into that saloon where he had been the life and soul of the learned world; and, though his speech was less fluent, he conversed with his physicians, his family, and the friends who aided them in their agonising cares. Among other anxious enquirers came M. Pasquier, whom he had seen on the memorable Tuesday; and he said to him, " Behold a very different person to the man of Tuesday — of Saturday. Nevertheless, I had great things still to do. All was ready in my head; after thirty years of labour

* " The nerves of the will are sick."
* A month before his illness, he had read a paper at the Institute upon a memoir of Scarpa's, on this distinction between the nerves of will, and those of sensibility.

and research; there remained but to write; and
now the hands fail, and carry with them the
head." M. Pasquier, almost too much distressed
to speak, attempted to express the interest universally felt for him; to which M. Cuvier replied,
" I like to think so; I have long laboured to
render myself worthy of it." In the evening,
fever showed itself and continued all night,
which produced great restlessness and desire for
change of posture; the bronchiæ then became
affected, and it was feared that the lungs would
soon follow. On Sunday morning the fever disappeared for a short time; consequently he slept;
but said, on waking, that his dreams had been
incoherent and agitated, and that he felt his
head would soon be disordered. At two o'clock
in the day, the accelerated respiration proved
that only a part of the lungs was in action; and
the physicians, willing to try every thing, proposed to cauterise the vertebræ of the neck : the
question, Had he right to die? rendered him
obedient to their wishes; but he was spared this
bodily torture, and leeches and cupping were all
to which they had recourse. During the application of the former, M. Cuvier observed, with
the greatest simplicity, that it was he who had

discovered that leeches possess red blood, alluding to one of his Memoirs, written in Normandy.
" The consummate master spoke of science for
the last time, by recalling one of the first steps
of the young naturalist." He had predicted
that the last cupping would hasten his departure; and, when raised from the posture necessary for this operation, he asked for a glass of
lemonade, with which to moisten his mouth.
After this attempt at refreshment, he gave the
rest to his daughter-in-law to drink, saying, it
was very delightful to see those he loved still
able to swallow. His respiration became more
and more rapid; he raised his head, and then
letting it fall, as if in meditation, he resigned
his great soul to its Creator without a struggle.

Those who entered afterwards, would have
thought that the beautiful old man, seated in
the arm-chair, by the fireplace, was asleep; and
would have walked softly across the room for
fear of disturbing him; so little did that calm
and noble countenance, that peaceful and benevolent mouth, indicate that death had laid his
icy hand upon them : but they had only to turn
to the despairing looks, the heart-rending grief,
or the mute anguish of those around, to be con-

vinced that all human efforts are unavailing, when Heaven recalls its own.*

The perfect disinterestedness of M. Cuvier's character, the remarkable liberality of his disposition, the sums he so delightedly bestowed on science, in a dearth of other proofs, would all be established by the moderate fortune he left to his family. After having filled such high offices in the state; after having executed, under the magnificent government of the empire, missions which a man thirsting after wealth would have turned to his pecuniary advantage; all the fortune he amassed amounted but to four thousand pounds sterling; his library had cost him a similar sum †; and he never hesitated procuring

---

* Germany lost her great Goëthe in this year. France, besides the above calamitous privation, was bereaved of Champollion, Casimir Perrier, and Abel Remusat, and Great Britain, of Sir Walter Scott and Sir John Leslie: the preceding year had been *her* greatest trial; for in it she was deprived of Sir Humphry Davy, Dr. Young, and Dr. Wollaston, &c.

† To the books purchased by himself were added those published at the expense of the Government, copies of which were always presented to him; and the numerous gifts he received from authors of all countries, who were universally anxious to pay him this mark of respect, even if their works did not treat of Natural History. Altogether amounted to more than nineteen thousand volumes, besides pamphlets, atlases, &c., and many of which contained his own notes. It was very desirable that this library should remain entire, for

any object of natural history at his own expense, original cost and freight included, from every quarter of the globe; not for himself, but to present it to the Museum : and if to these be added his hospitality, and his generous assist-

---

the use of students ; and such being Madame Cuvier's wish, the legatees, consisting of M. F. Cuvier, his son, M. Valenciennes, and M. Laurillard, accepted the value of their portions as mere books, and the Government agreed to purchase the whole. The sum was voted at the same time as Madame Cuvier's pension; and much is it to be regretted that the value of books has of late years so much diminished in France : however, it is much more vexatious, that no building can be found to contain this collection, where it might be consulted in its entire state by the public ; and it is therefore to be divided between the Schools of Law and Medicine, the Normal School, and the Jardin des Plantes, where many volumes will enter as duplicates. The apartments in which these treasures were contained, were a continuation of M. Cuvier's own dwelling, and had been originally used for the forage of the menagerie. On this being removed to the building called the Rotonde, Baron Cuvier asked permission of the Board of Administration of the Jardin, to take these granaries into his own hands, and convert them, at his own expense, into a suite of rooms. This cost him 1640*l.*, which gave him a right to ask for a dwelling for his family after his death ; a right which was graciously confirmed by his present Majesty. In these rooms the great savant carried on his vast labours and meditations, working in each according to the subject on which he was employed: they made his house appear large ; but, in reality, the habitable part of it was scarcely of sufficient extent for his comfort, when it is considered how many visitors he was there obliged, by his places, to entertain.

ance to others, the small amount of the property
he left behind him may be easily accounted for.
He desired to be buried without ceremony, in
the cemetery of Père la Chaise, under the tomb-
stone which covered his daughter; but it was
not possible for such a man to die without much
public manifestation of respect at the last sad
ceremony. The funeral procession was followed
by a deputation from the Council of State, presided by the Keeper of the Seals; also from the
Academies of Sciences, of Inscriptions, of Medicine, of France; by members of the two Chambers, the Ecole Polytechnique, &c. The earthly
remains were alternately borne by pupils from
the laboratories of the Jardin des Plantes, from
the Schools d'Urfort, of Law, and of Medicine,
and first taken to the Protestant Church in the
Rue des Billettes. The pall was supported by
M. Pasquier, president of the Chamber of Peers;
M. Devaux, counsellor of state; M. Arago, secretary of the Academy of Natural Sciences;
and M. Villemain, vice-president of the Royal
Council of Public Instruction. Different members of the learned and legislative bodies, each
pronounced a funeral discourse over the grave,
according to the usual custom of the country.
A monumental statue is to be erected in the

Jardin des Plantes, another at Montbéliard, the
size and materials of which depend on the
amount of the subscriptions. The King has
also ordered a marble bust, by M. Pradhier, to
be placed in the Institute; and another to be
placed in the Galleries of Anatomy, by M. David.
M. Cuvier is succeeded by Baron Dupin (the
elder) at the Académie Française, and by Dr.
Dulong * at the Académie des Sciences. M. de
Blainville is appointed professor of comparative
anatomy at the Jardin des Plantes. Many of
his places remain unfilled, as if those, who would
otherwise be candidates, were afraid of the contest. This one man held them all; rigidly performed all their duties; carried his benevolent
and enlightened principles with him into all his
employments; scorned no detail which could
bear upon their improvement; saw, in one
glance, the influence which their progress would
have over society at large; and yet, while his
mind was filled with these great and general
views, never, for one instant, forgot that which
belonged to his character as a father, a husband,
a brother, and a friend; or that he had fellow
creatures who needed his assistance. His public

* Since writing the above, M. Dulong has resigned his
secretaryship, on account of his health.

employments are now separated; and the occu-
piers may think themselves happy, if they can,
in their solitary succession, in some degree attain the perfection which stamped his combined
career. The death of such a man, at such a period
of his labours, and at such a moment, scarcely
seems to come within the common routine of
mortality, but to have been the result of a special and chastening mandate from Heaven.

CHRONOLOGICAL LIST

OF

THE PRINCIPAL EVENTS

OF

THE BARON CUVIER'S LIFE.

A. D.

1769. (*August* 23.) Born.

1779. Entered the Gymnase of Montbéliard.

1784. (*May* 4.) Entered the Académie Caroline, in the University of Stuttgardt.

1788. Left Stuttgardt to return to Montbéliard.

    Entered as tutor into the family of Count d'Hericy, in Normandy.

1793. Death of M. Cuvier's mother.

1795. (*Spring.*) Came to Paris.

    Appointed Membre de la Commission des Arts.

    Appointed Professor at the Central School of the Panthéon.

    (*July.*) Made assistant to M. Mertrud, and entered the Jardin des Plantes ; sent for his father and brother ; commenced the Gallerie d'Anatomie comparée.

    (*December.*) Opened his first course of lectures, at the Jardin des Plantes, on Comparative Anatomy.

CHRONOLOGICAL LIST

OF

THE PUBLISHED WORKS

OF

THE BARON CUVIER.

A. D.

1792. Mémoire sur l'Anatomie de la Patelle.

1795. Mémoire sur le Larynx inférieur des Oiseaux. (Magasin Encyclopédique.)

Mémoire sur l'Anatomie du grand Limaçon. (Helix Pomatia *Lin.*)

Notice ou Mémoire sur la Circulation dans les Animaux à sang blanc.

Mémoire sur une nouvelle Division des Mammifères. (Magasin Encyclopédique.)

Mémoire sur une nouvelle Distribution, en six Classes, des Animaux à sang blanc.

Mémoire sur la Structure des Mollusques, et de leur Division en Ordres.

A. D.
1796.   Made a Member of the National Institute.

1798.   Proposal made to M. Cuvier, by Count Berthollet, to accompany the expedition to Egypt; which offer was refused.

A. D.

1796. Mémoire sur le Squelette d'une très grande Espèce de Quadrupède inconnue (Megalonix).

Mémoire sur les Têtes d'Ours Fossiles, des Cavernes de Gailenreuth.

Mémoire sur un Squelette Fossile (Megatherium) trouvé sur les Bords du Rio de la Plata.

Mémoire sur l'Organe de l'Oïüe dans les Cetacés.

Mémoire sur un nouveau Genre de Mollusque (Phyllidia).

1797. Mémoire sur l'Animal des Lingules.

Note sur l'Anatomie des Ascidies.

Note sur les différentes Espèces de Rhinoceros.

Note sur les Narines des Cetacés.

Note sur les Rates du Marsouin.

Note sur une nouvelle Espèce de Guêpe Cartonnière Elogé Historique de Riche.

Mémoire sur la manière dont se fait la Nutrition dans les Insectes.

1798. Tableau Elémentaire de l'Histoire Naturelle des Animaux.

Mémoire sur les Organes de la Voix dans les Oiseaux.

Mémoire sur les Ossemens Fossiles des Quadrupèdes. Ici sont indiqués l'Elephant, le Mastodonte d'Amerique et d'Europe, l'Hippopotame, le Rhinoceros à crane allongé, le Tapir gigantesque, le Megatherium, l'Ours des Cavernes, un Animal carnassier de Montmartre (reconnu plus tard pour être un Pachyderma), l'Animal de Monti, que M. Cuvier croyait un Mastodonte, l'Elan d'Islande, qu'il croyait alors, sur les rapports de Faujas, exister à Maestricht, deux Espèces de Bœufs de Sibérie, deux Cerfs des Tourbières de la Somme.

Mémoire sur les Vaisseaux sanguins des Sangsues, et sur la couleur rouge du Fluide qu'y est contenu. (Celle-ci est la découverte sur laquelle repose l'établissement de la classe des Vers à sang rouge.)

Mémoire sur les Ossemens qui se trouvent dans les

A. D.

1800. Appointed Professor at the Collège de France, on
       which M. Cuvier resigned the chair at the Central
       School of the Panthéon.
          Elected Secretary to the Class of Physical and Ma-
       thematical Sciences of the Institute.

1802. Named one of the six Inspector-Generals of Educa-
       cation (Études).

A. D.

Gypses de Montmartre. (Ici M. Cuvier rectifie son Mémoire précédent, et annonce avoir reconnu trois espèces distinctes de Pachydermes.)

1799. Notice Biographique sur Bruguières.

Mémoire sur les Différences des Cerveaux, considerés dans tous les Animaux à Sang rouge.

Mémoire sur l'Organisation de quelques Meduses (Rhyzostome bleu).

1800. Mémoire sur les Tapirs Fossiles de France.

Mémoire sur le Siren Lacertina.

Mémoire sur un nouveau Genre des Quadrupèdes édentés, nommés Ornithorynchus paradoxus, décrit par Blumenbach (extrait par M. Cuvier).

Mémoire sur l'Ibis des anciens Égyptiens.

Mémoire sur les Ornitholithes de Montmartre.

Addition à l'Article des Quadrupèdes Fossiles, où est indiqué l'Anoplotherium, et une Espèce du même Genre, de la taille d'un Hérisson.

Mémoire sur une nouvelle Espèce de Quadrupède Fossile, du Genre de l'Hippopotame.

Tomes I. et II. des Leçons de l'Anatomie comparée.

Éloge Historique de Daubenton.

Éloge Historique de Lemonnier.

1801. Mémoire sur une nouvelle Espèce de Crocodile Fossile, des Environs de Honfleur.

Note sur des nouvelles découvertes d'Os Fossiles. (Il s'agit des Crocodiles de Honfleur, d'Altorf en Franconie, de Provins, Département de l'Orme.) Ici M. Cuvier annonce le découverte d'un septième animal dans le gypse de Montmartre, un Carnassier (Canis).

Mémoire sur les Dents des Poissons.

Éloge Historique de l'Héritier.

Éloge Historique de Gilbert.

1802. M. Cuvier commença les Analyses des Travaux de l'Institut, qui étaient continués jusqu'à sa mort.

A. D.

1802. Went to Marseilles, &c. to found the Royal Colleges.

1803. Made perpetual Secretary to the Class of Physical
and Mathematical Sciences of the Institute.
Resigned Inspector-generalship of Education.
Married to Madame Duvaucel.

A. D.

1802. Mémoire sur l'Animal de Lingule, l'Animal de Bullæa aperta, et celui de Clio Borealis.

Mémoire sur le Genre Tritonia, avec la Description d'une Espèce nouvelle.

Éloge Historique de Jean Darcet.

Extrait d'un Mémoire sur les Vers qui ont le sang rouge. Ici M. Cuvier annonce, que la plupart des Vers marins ont le sang rouge, ainsi que les Lombrics ; et donne la description du système circulatoire dans l'Arénicole, ou Lombric Marin.

Extrait de la Description de l'Anatomie de l'Ornithorynchus p. par Home.

Mémoire sur les Serpules.

Articles Abdomen, Absorption, Accouplement, Acéphales, Actinie, pour la Dictionnaire des Sciences Naturelles.

1803. Mémoire sur le Genre Aplysia.

Mémoire sur les Écrevisses connues des Anciens, &c.

Notice sur l'Établissement de la Collection d'Anatomie comparée du Museum.

Description Ostéologique du Rhinoceros Unicorne.

Description Ostéologique du Tapir.

Description Ostéologique du Daman.

Mémoire sur les Espèces des Animaux dont proviennent les Os Fossiles répandus dans la Pierre à Plâtre des Environs de Paris.

Premier Mémoire — Restitution de la Tête.

Second Mémoire — Examen des Dents.

Troisième Mémoire — Restitution des Pieds.

Mémoire sur les Os Fossiles des Environs de Paris.

Article Historique sur les Collections de l'Histoire Naturelle.

Note sur l'Anatomie de quelques Aplysies, observés pendant un séjour à Marseille.

Mémoire sur la Pennatula Cynomorium, et sur les Coraux en general, ——. montre que la Pennatula

Z

A. D.

1804.   Eldest son born, and died.

A . D.

Cyn. est composée des plusieurs Animaux, avec une seule volonté, ce qu'on déduit de leurs mouvemens, qu'il y a unité de nutrition, et qu'on peut la regarder comme un seul animal à plusieurs bouches.   M. Cuvier étend la même conclusion aux Zoophytes fixés, quoiqu'ils différent essentiellement par l'absence du mouvement.

1804.   Article Bec, pour la Dictionnaire des Sciences Naturelles.

Recherches d'Anatomie comparée sur les Dents.

Notice sur un Squelette Fossile, trouvé à Pantin, dans le Gypse (Paleotherium minus).

Mémoire sur l'Hyale, sur un nouveau Genre des Mollusques nus, intermédiaire entre l'Hyale et le Clio, et l'établissement d'un nouvel Ordre dans la Classe des Mollusques.

Mémoire sur l'Hippopotame et son Ostéologie.

Mémoire sur les Thalides, et sur les Biphores.

Mémoire sur le Genre Doris.

1805.   Articles Bœuf, Bois, Branchie, pour la Dictionnaire des Sciences Naturelles.

Trois derniers Volumes des Leçons de l'Anatomie comparée.

Éloge Historique de Priestley.

Mémoire sur les Animaux auxquels appartenaient les Pierres dites Nummulaires, ou Lenticulaires, et sur ceux du Corne d'Ammon. (M. Cuvier attribue les Nummulaires concentriques à des osselets intérieurs d'un Zoophyte, voisin des Porpites.)

Extraits des Mémoires sur le Clio Borealis, l'Hyale, le Pneumoderme.

Suite des Mémoires sur les Tritonia, Doris, Aplysia, Onchidium, Bullæa.

Suite des Mémoires précédents.

Suite des Mémoires sur la Phyllidia et le Pleurobranchus.

A. D.

A. D.

1806. Éloge Historique de Cels.

Mémoire sur les Os Fossiles trouvés en divers endroits de la France, et plus ou moins semblables à ceux du Paleotherium.

Mémoire sur la Scyllée, l'Eolide, et le Glaucus, avec des Additions au Mémoire sur la Tritonia.

Mémoire sur l'Onchidium Peronii.

Additions à l'Article sur les Ossemens Fossiles des Tapirs.

Additions à l'Article sur l'Hippopotame.

Mémoire sur les Ossemens Fossiles d'Hippopotame.

Mémoire sur la Phyllidie et sur le Pleurobranche.

Mémoire sur le Sarigue Fossile des Gypses de Paris.

Mémoire sur le Megalonyx.

Mémoire sur le Megatherium.

Mémoire sur la Dolabelle.

Mémoire sur les Rhinoceros Fossiles.

Mémoire sur le Limaçon et le Colimaçon.

Mémoire sur les Ours des Cavernes d'Allemagne.

1807. Éloge Historique de Michel Adanson.

Mémoire sur les Elephans vivans et Fossiles.

Mémoire sur le Grand Mastodonte.

Mémoire sur les autres Espèces de Mastodonte.

Résumé général de l'Histoire des Ossemens Fossiles, de Pachydermes, des Terreins Meubles et d'Alluvion.

Mémoire sur les Ossemens Fossiles des Environs de Paris. Les Phalanges.

Mémoire sur les Os des Extrémités.

Mémoire sur les Os longs des Extrémités.

Mémoire sur les Extrémités antérieures.

Mémoire sur les Omoplates et les Bassins.

Description de deux Squelettes presque entiers de l'Anoplotherium commun.

Mémoire sur les Ornitholithes de la Pierre à Plâtre de Paris.

Mémoire sur les Carnassiers (autres que l'Ours) des Cavernes.

1808.   Appointed Counsellor to the University.

1809. ⎫ Charged with the organisation of the Academies of
1810. ⎭   the Italian States.

A. D.

1807. Mémoire sur les différentes Espèces de Crocodiles vivans.

Mémoire sur quelques Ossemens de Carnassiers dans les Carrières à Plâtre de Paris.

Rapport à la Classe des Sciences Physiques et Mathématiques de l'Institut, sur l'Écrit de M. Adams sur l'Elephant Fossile.

Rapport sur un Mémoire de M. Decandolle, intitulé " Tableau de la Nutrition de Végétaux."

Rapport sur un ouvrage manuscrit de M. André, ci-devant connu sous le Nom de Père Chrysologue de Gy, lequel ouvrage (sur la Géologie) est intitulé " Théorie de la Surface actuelle de la Terre."

1808. Rapport Historique sur les Progrès des Sciences Naturelles, depuis 1789, &c.

Eloge Historique de Broussonnet.

Mémoire sur l'Ianthine et la Phasianelle.

Mémoire sur l'Helix vivipara.

Rapport sur un Mémoire de MM. Gall et Spurzheim.

Mémoire sur le Buccinum undatum.

Essai sur la Géographie Minéralogique des Environs de Paris (avec M. Brongniart).

Observations sur les Ossemens Fossiles des Crocodiles, sur ceux des Environs de Honfleur, du Havre, et de Thuringie.

Observations sur le grand Animal Fossile de Maestricht.

Mémoire sur le Genre Thétys.

Suite des Recherches sur les Ossemens Fossiles des Environs de Paris.

Mémoire sur les Os des Ruminans des Terreins Meubles.

1809. Mémoire sur les Brèches Osseuses de Gibraltar.

Mémoire sur l'Ostéologie du Lamantin et du Dugong.

1811. Charged with the organisation of the Academies of
Holland.
Received the title of Chevalier.

1812. Death of Mademoiselle Anne Cuvier.

1813. Death of George Cuvier, jun.
M. Cuvier sent to Rome, to organise the University
there.
Named Maître des Requêtes.
Ordered to make a list of books for the King of
Rome, with an intention that M. Cuvier should super-
intend his education.
Made Commissaire Impériale Extraordinaire, and

A. D.

1809.   Mémoire sur quelques Quadrupèdes Fossiles des Schistes Calcaires.

Mémoire sur les Ossemens Fossiles des Chevaux et des Sangliers.

Supplément au Mémoire sur les Ornitholithes de Paris.

Mémoire sur les Rongeurs Fossiles des Tourbières, et sur quelques autres Rongeurs, renfermés dans les Schistes.

Mémoire sur les Espèces vivantes des grands Chats.

Rapport sur le Mémoire de Delaroche, sur la Vessie Aërienne des Poissons.

Mémoire sur les Ossemens Fossiles des Tortues.

1810.   Mémoire sur les Acères.

Mémoire sur les Reptiles et les Poissons des Gypses de Paris.

Éloges Historiques de Bonnet et de De Saussure.

1811.   Recherches sur les Ossemens Fossiles (grand ouvrage en quatre volumes in 4to.).

Éloge Historique de Fourcroy.

Rapport sur un Mémoire de M. Jacobson, intitulé " Description Anatomique des Organes observés dans les Mammifères."

1812.   Article Animal, pour la Dictionnaire des Sciences Médicales.

Éloge Historique de Dessesserts.

Éloge Historique de Cavendish.

1813.   Articles Azygos, Caverneux, pour la Dictionnaire des Sciences Naturelles.

Rapport sur des Cetacés echoués sur les Côtes de France, le 7 Janvier, 1812.

Mémoire sur un nouveau Rapprochement à établir entre les Classes qui composent le Règne Animal.

Mémoire sur la Composition de la Tête Osseuse dans les Animaux Vertebrés.

Éloge Historique de Pallas.

sent to the left bank of the Rhine, in order to take the
steps necessary for opposing the invasion of France.

1814.   Named Counsellor of State by Napoleon.
Named Counsellor of State by Louis XVIII.
(*September.*)   First officiated as Commissaire du Roi,
to which he was repeatedly called at various periods of
his life.
Named Chancellor of the University.

1815.   Procured ameliorations of the Criminal Laws, and in
the Prévôtal Courts.

A. D.

1813. Mémoire sur le Lophote Giorna.

Article Dent, pour la Dictionnaire des Sciences Médicales.

1815. Éloges Historiques de Parmentier, et du Comte de Rumford.

Mémoire sur l'Aigle au Maigre.

Mémoire sur la Composition de la Machoire inférieure des Poissons.

Observations et Recherches Critiques sur les Poissons de la Mediterranée.

Suite du même.

Suite du même.

Suite du même.

Mémoire sur les Ascidies.

Mémoire sur les Anatifes et Balanes.

Rapport sur deux Mémoires de M. Savigny, intitulés " Observations sur les Alcyons, (à la suite des Mémoires sur les Animaux sans Vertèbres de Savigny, 2me partie, page 67.).

1816. Réflexions sur la Marche actuelle des Sciences, &c.

Éloge Historique d'Olivier.

1817. Éloge Historique de Tenon.

Articles Cartilage, Cerveau, pour la Dictionnaire des Sciences Naturelles.

Seconde Édition des Recherches sur les Ossemens Fossiles, en cinq volumes in 4to.

Le Règne Animal, en quatre volumes in 8vo.

Rapport sur un Mémoire de M. Dutrochet, intitulé " Recherches sur les Enveloppes du Fœtus."

Mémoire sur les Œufs des Quadrupèdes.

Mémoire sur la Venus Hottentote.

A. D.

1818.   Offered the Ministry of the Interior; which offer was refused.
     First journey to England.
     Elected Member of the Académie Française.

1819. (*September* 13.) Named temporary Grand Master to the University.
     Appointed President of the Comité de l'Intérieur.
     Created a Baron.
1820. (*December* 21.)   Resigned Grand Mastership.

1821. (*July* 31.) Appointed temporary Grand Master to the University.

1822.   (*June* 1.) Resigned Grand Mastership.
     Made Grand Master of the Faculties of Protestant Theology.

1824.   Officiated as one of the Presidents of the Council of State, at the coronation of Charles X.
     Made Grand Officier de la Légion d'Honneur.
     Made Commander of the Order of the Crown, by the King of Würtemburg.

A. D.

1818. Article Hymen, pour la Dictionnaire des Sciences Médicales.

    Éloge Historique de Werner.

    Éloge Historique de Desmarets.

    Mémoire sur le Genre Chironectes.

    Mémoire sur les Diodons.

    Mémoire sur le Genre Myletus.

    Discours sur la Réception de M. Cuvier à l'Académie Française.

1819. Mémoire sur les Poissons du Genre Hydrocyn.

1820. Éloge Historique de M. de Beauvois.

    Mémoire sur le Meleagris Ocellata.

1821. Rapport sur une Mémoire de M. Audouin, intitulé " Recherches Anatomiques sur le Thorax des Animaux Articulés, et celui des Insectes en particulier. (Annales des Sciences Physiques de Bruxelles, vii. Journal de Physiologie Expérimentale, i.)

    Éloge Historique de Sir Joseph Banks.

1822. Rapport sur un Mémoire de M. Flourens, sur le Système Nerveux.

    Éloge Historique de M. Duhamel.

    Discours Funèbre de M. Vanspaendonck.

    Discours Funèbre de M. Délambre.

1823. Éloge Historique de Haüy.

1824. Mémoire sur une altération singulière de quelques Têtes Humaines.

    Mémoire sur le Bradypus tridactylus.

    Rapport sur l'État de l'Histoire Naturelle, et sur ses accroissemens.

    Éloge Historique du Comte Berthollet.

    Éloge Historique de Richard.

1825. Article Nature, pour la Dictionnaire des Sciences Naturelles.

    Seconde Édition du Discours Préliminaire des Re-

A. D.

1827. (*June* 14.) Appointed Censor of the Press; which appointment was instantly refused.

    Charged with the government of all the non-Catholic religions.

1828. (*September* 28.) Death of Mademoiselle Clementine Cuvier.

1830.   Resumed lectures at the Collége de France.

    Paid a second visit to England.

1832.   Created a Peer.

    (*May.*) Appointed President to the entire Council of State.

    (*May* 13.) Death.

A. D.

cherches sur les Ossemens Fossiles, appellé " Discours sur les Révolutions de la Surface du Globe," in 8vo.

1825. Discours sur la Distribution des Prix de Vertu.

Éloge Historique de Thouin.

1826. Éloge Historique du Comte de Lacépéde.

Rapport sur les Principaux Changemens éprouvés par les Theories Chimiques.

Édition in 4to. du " Discours sur les Révolutions du Globe."

1827. Éloges Historiques de MM. Hallé, Corvisart, et Pinel.

Éloge Historique de M. Fabbroni.

Mémoire sur le Canard Pie de la Nouvelle Hollande.

1828. Volumes I. et II. du grand Ouvrage sur l'Ichthyologie.

Éloge Historique de Ramond.

Caii Plinii Secundi, Libri de Animalibus,————— Notas et Excursus Zoologici Argumenti adjecit, G. Cuvier (traduits en 1831).

Rapport fait à l'Institut sur un Mémoire de M. Adolphe Brongniart, intitulé " Considérations générales de la Nature de la Végétation qui couvrait la Surface de la Terre, aux divers périodes de la formation dé son Écorce."

1829. Seconde Édition du Règne Animal, en 5 tom. in 8vo.

Volumes III. et IV. sur l'Ichthyologie.

Éloge Historique de M. Bosc.

1830. Volumes V. et VI. sur l'Ichthyologie.

Éloge Historique de Sir Humphry Davy.

Éloge Historique de Vauquelin.

1831. Volumes VII. et VIII. sur l'Ichthyologie.

1832. Éloge Historique de Lamarck.

Et en outre plusieurs Rapports sur les Collections rapportés par les Voyageurs, tels que les Collections de MM. Quoy et Gaimard, Lesson et Garrot, Dussumier, &c. &c.

THE END.

ERRATUM.

Page 19. note, for " Baron Pasquier " read " The Baron de H ———."

LONDON:
Printed by A. SPOTTISWOODE,
New-Street-Square.

Printed in the United States
By Bookmasters